Technology-Based Inquiry for Middle School

An NSTA Press Journals Collection

Technology-Based Inquiry for Middle School

An NSTA Press Journals Collection

Edwin P. Christmann, Editor

An NSTA Press Journals Collection

Arlington, Virginia

Claire Reinburg, Director
Judy Cusick, Senior Editor
Andrew Cocke, Associate Editor
Betty Smith, Associate Editor
Robin Allan, Book Acquisitions Coordinator

PRINTING AND PRODUCTION Catherine Lorrain, Director
Nguyet Tran, Assistant Production Manager
Jack Parker, Electronic Prepress Technician
Will Thomas, Jr., Cover and Book Design

NATIONAL SCIENCE TEACHERS ASSOCIATION
Gerald F. Wheeler, Executive Director
David Beacom, Publisher

09 08 07 06 4 3 2 1

LIBRARY OF CONGRESS CATALOGING-IN-PUBLICATION DATA

Technology-based inquiry for middle school: an NSTA Press journals collection / Edwin P. Christmann, editor.
p. cm.
Includes index.
ISBN 0-87355-266-0
1. Science--Study and teaching (Middle school)--Technological innovations. 2. Science--Computer-assisted instruction. 3. Educational technology. I. Christmann, Edwin P., 1966-
Q181.T3544 2006
507'.12--dc22
2006001302

Contents

Section I: Scientific Inquiry

Section II: Physical Science

Section III: Earth and Space Science

Section IV: Life Sciences

Section V: General Science and Technology Applications

Additional Materials Available Online

Technology constantly evolves as new discoveries are made and new uses are found. To supplement the chapters in this book, I've created a web page with additional information on each chapter. There you'll find up-to-date information, including chapter reviews, outlines, sample test questions, and activities. You can download PowerPoint presentations for teaching the text as well.

Please visit *http://srufaculty.sru.edu/edwin.christmann/epc2.htm.*

Scientific Inquiry

"The aim of natural science is not simply to accept the statements of others, but to investigate the causes that are at work in nature."

—Albertus Magnus

EDWIN P. CHRISTMANN

A brief history

In 1972 Karl Popper traced the beginnings of the scientific method to the turn of the sixth and fifth centuries B.C., in ancient Greece. During this time the Greeks tried to understand or explain the structure of the Universe in terms of the story of its origin. It was not until the 13th century, however, that Albert the Great (i.e., Albertus Magnus, circa 1197 to 1280 AD), the prolific Dominican Friar and professor, wrote 36 volumes on what was then known as "natural philosophy." Subsequently he is known as the father of the natural sciences, which are now divided into physics, geology, astronomy, chemistry, and biology.

During the medieval era, science was not the process of inquiry that it is today. Therefore, as an early scientist, Albert the Great relied on his encyclopedic scientific knowledge, which he synthesized from Aristotle's Greek texts and the Arab writings. It is with Albert the Great that the earliest scientific experiments are documented with students in his laboratory at the University of Cologne. As an experimenter, Albert the Great built up a collection of plants, insects, and chemical compounds, laying the groundwork for later scientific inquiry in his laboratory. Kovich and Shahan (1980) verify that during Galileo's professorate at the University of Pisa, his notebooks mention Albert the Great 23 times in his logical and physical questions. Clearly, it is around this era during the 15th century, in the time of Galileo, when science transforms into a process of inquiry.

Subsequently, technological advances led to the development of experimental tools, an advancement that catapulted scientific knowledge into an ongoing process of scientific investigation. No longer did an alchemist work as a mere craftsman in that the modern era of science is a field harnessed by technology, steered by the scientific method, and fueled by the process of scientific inquiry. Undoubtedly, the applications of the latest technologies (e.g., microcomputers, internet, and calculators) have increased the rate at which scientific problems are investigated and solved. There is little doubt that these technological advances have improved the quality of life for the majority of people throughout the world. Thus, students should be familiar with the latest technologies that are used in the process of scientific inquiry, as well as have rich experiences in science classrooms in middle schools throughout the United States.

What is science?

Trefil and Hazen (2004) explain that, based on experiments and observation, science is a way of knowing that answers questions about the natural world that surrounds us. Subsequently, science is based on verifiable facts about physical phenomena. According to the tenets of the National Science Education Standards (1996), students should be guided by the following principles when studying science:

- Science is for all students.
- Learning science is an active process.
- School science reflects the intellectual and cultural traditions that characterize the practice of contemporary science.
- Improving science education is part of systematic education reform.

Science is everywhere around us. Science is involved with the water that we drink, the food that we cook, the bicycles that we ride, and the stars in the night sky. Most important, however, is that science is fun. The activities in this book integrate some of the latest technologies into classroom activities, which will hopefully provide students with some of the excitement that scientists have experienced through the joy of conducting their own scientific experiments. Most important, however, is that students gain an understanding of some of the key terminology that is used by scientists. Below are some very important concepts that are prerequisite to an in-depth understanding of scientific inquiry.

Scientific hypotheses

Based on observations, science is a collection of knowledge about nature of the physical world. Hence, scientists make hypotheses (educated guesses) in attempts to explain observations by testing hypotheses through experiments. Therefore, scientific questions can be tested and verified through experiments, resulting in new knowledge that can be built upon by future generations.

Scientific theories

Scientific theories are tentative detailed explanations and descriptions of the world that cover a relatively large number of phenomena. Theories offer the scientific community testable observations that are predictable and useful for further investigation. Some examples of scientific theories are relativity, evolution, and plate tectonics. It is important to emphasize, however, that theories are unconfirmed and may be modified or even discarded with new scientific findings. Hence, scientific theories help to expand the body of scientific knowledge, which is constantly developing, changing, and contrary to popular opinion, is never absolute.

Scientific laws

Scientific laws are based on large amounts of scientific data and can be summarized by a brief statement. For example, Newton's First Law of Motion states, "Every object either remains at rest or in continuous motion with constant speed unless acted upon by external forces." A scientific law is subjected to rigorous testing by a variety of experiments that are replicated several times. A valid scientific law can predict natural phenomena with great precision. For example, Newton's Second Law of Motion states, "The rate of change of momentum of a moving body is proportional to and in the same direction as the force acting on it." For example, if you kick a football, its path through the air is not a straight line; due to gravity the football curves toward the Earth. Many times laws are expressed mathematically. For example Newton's Second Law of Motion can be expressed algebraically as $F = ma$.

Scientific models

A scientific model can be used to help explain phenomena by revealing ideas about a complicated system from a simple system. For example, Figure 1 is a NASA Rainfall Model that forecasts weather observations with a computer-based

Figure 1 NASA Rainfall Model

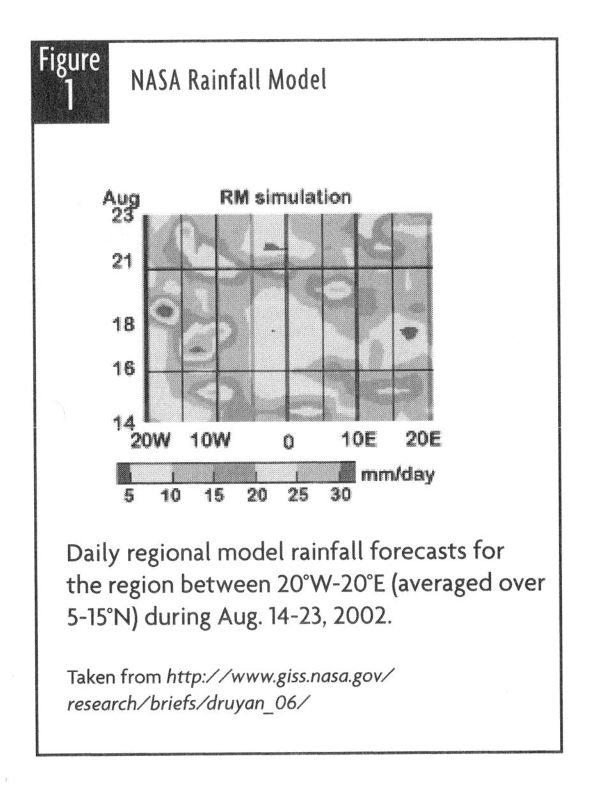

Daily regional model rainfall forecasts for the region between 20°W-20°E (averaged over 5-15°N) during Aug. 14-23, 2002.

Taken from *http://www.giss.nasa.gov/research/briefs/druyan_06/*

Figure 2 Human Torso Model

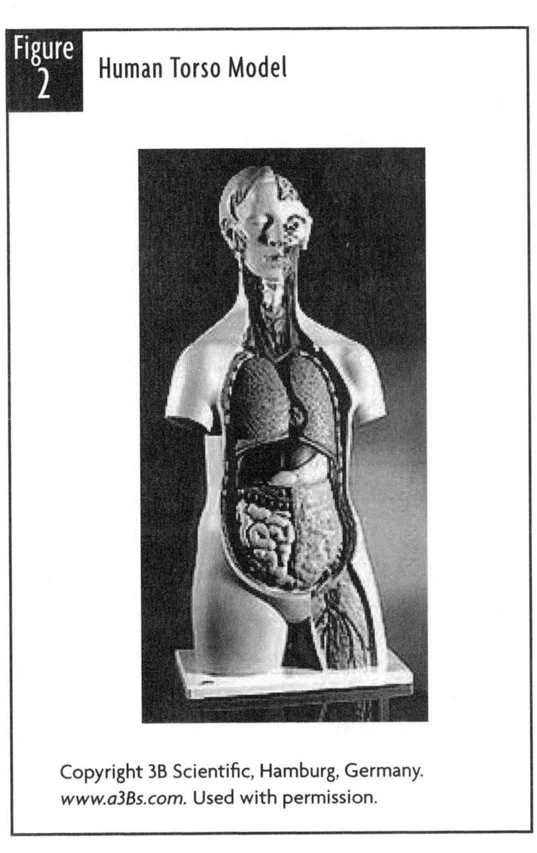

Copyright 3B Scientific, Hamburg, Germany. *www.a3Bs.com.* Used with permission.

model. The computer-based model uses predictions based on past observations to fill in the data sets. Some other examples of scientific models are mixing baking soda and vinegar to simulate volcano eruptions, using marbles to represent gases, and examining a model of a human torso (see Figure 2). Subsequently, scientific models can give visual representations to represent something that cannot readily be seen.

Scientific inquiry

The National Research Council (1996) emphasizes "scientific inquiry" as a way for students to be engaged with conceptualized questions that seek explanations of the world surrounding them. In essence, learning through scientific inquiry gives students an opportunity to investigate problems methodologically; similar to the procedures used by scientists. To simplify the process of scientific inquiry, Figure 3 shows how an observation of a natural phenomenon is questioned and tested through a sequential process. Based on a hypothesis (a plausible explanation), an experiment is conducted, which culminates into a conclusion that is based on whether or not the hypothesis is rejected.

Again, it is important to emphasize that scientific inquiry is an ongoing process that can result in replicated experiments so that the scientific community have valid and reliable findings. As discussed earlier, this process of replication draws on previous experiments and uses known scientific theories, laws, and models. Therefore, as presented in Figure 4, scientific inquiry is a cyclical process that, based on a natural phenomenon, begins with a question. For example, if we are interested in the air temperature at a specific location, we can set up an experiment to take temperature readings. Subsequently, by recording hourly temperature readings at the same location, we can compare readings taken in September to

Figure 3 Simplest Form of Scientific Inquiry

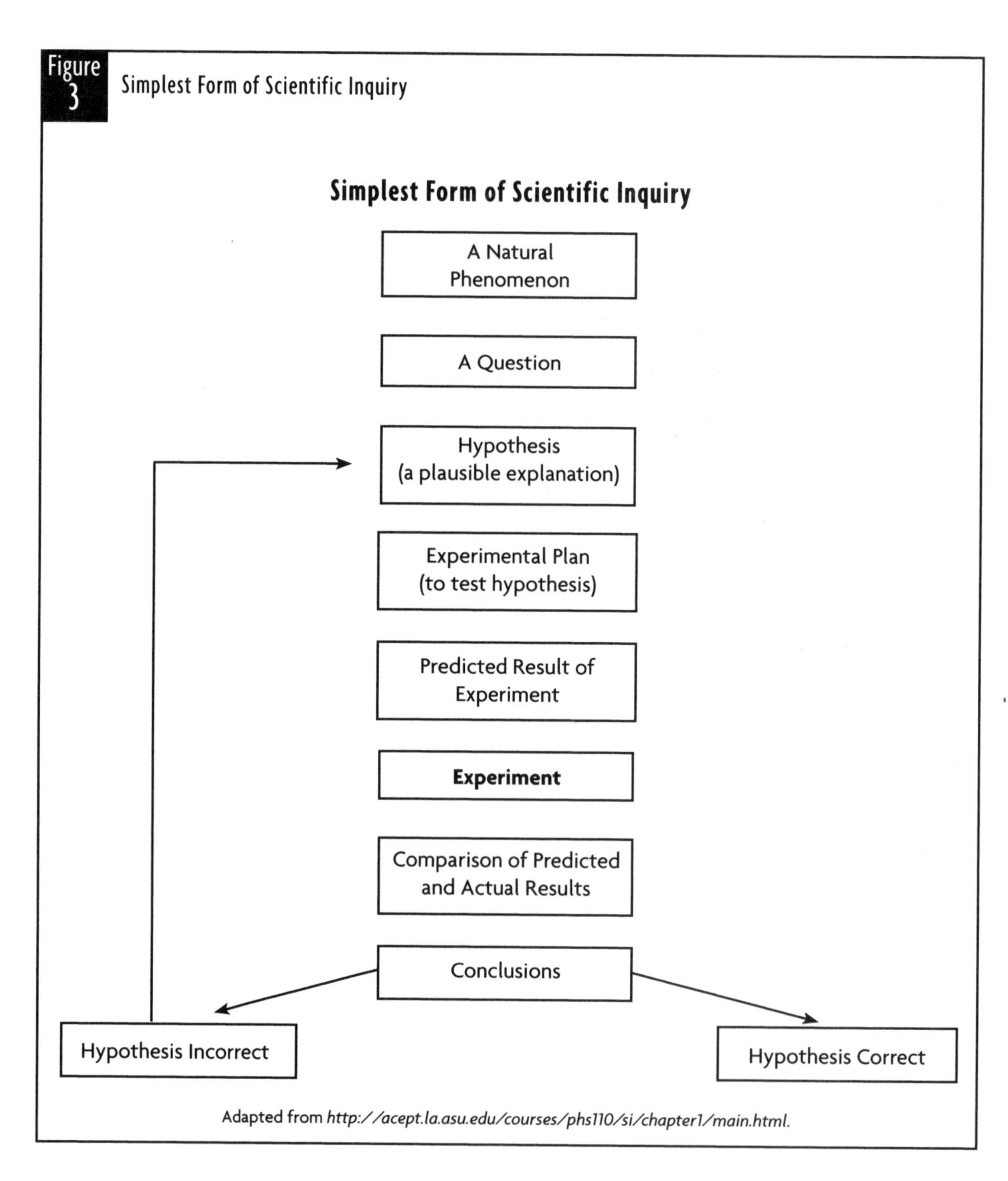

Adapted from *http://acept.la.asu.edu/courses/phs110/si/chapter1/main.html.*

those taken in January and examine the variation between the results. Once the results have been gathered, a computer spreadsheet model with graphics can be used to show differences in factors that have caused temperature differences (e.g., direct sunlight time). Thus, models can be used to present additional experiments. Clearly, scientific inquiry is essential for the middle school science student in that this process makes connections with prior scientific knowledge.

The middle school science student

Most middle school science teachers are aware that students in grades 6 through 8 are developing

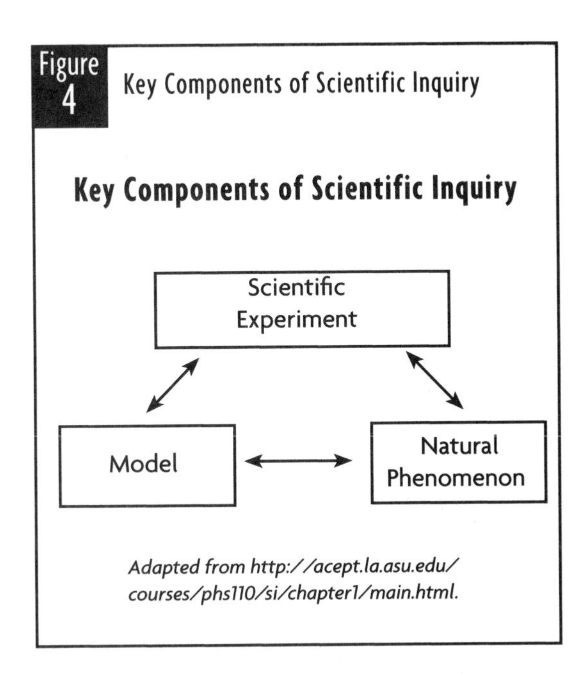

Figure 4 Key Components of Scientific Inquiry

Adapted from http://acept.la.asu.edu/courses/phs110/si/chapter1/main.html.

abstract and logical thinking skills. Jean Piaget, the famous developmental psychologist, explained that middle school age children are moving through a developmental period where they are able to generate abstract propositions and multiple hypotheses. At this stage, which is known as Formal Operations, thinking becomes less tied to concrete reality and more reliant on rational judgments. Simultaneously, the world surrounding them is changing in a seemingly random, yet predictable, pattern. For example, children now are observing objects moving through the sky, and noticing, for example, that the Earth's materials have different physical and chemical properties. Perhaps more obvious to middle school students, however, are the changes that surround them from day to day and over the seasons. Subsequently, this stage in human maturation is ideal for middle school learners to be engaged in learning activities that are scientific and inquiry-rich.

Technology-based inquiry

The transition from using simple "found tools" to making tools and keeping them for future use is the essence of the intellectual and social evolution of the human species (Devore, 1980). As discussed earlier, the use of technologies has increased the rate at which scientific problems are solved. For instance, John Harrison, a 19th Century cabinetmaker, solved the problem of longitude by using some of the tools of his era (i.e., mechanical clocks and basic astronomy) to solve one of the most difficult problems of his age. Harrison's technological solution to the problem of longitude was to compare a clock's time at the prime meridian to the time of the unknown position at noon at that unknown point. Subsequently, the time at noon of the unknown position was compared to a mechanical clock (i.e., a chronometer) carried on a ship. Hence, the mechanical clock needed to provide a precise time measurement for the prime meridian as a comparison to the noontime reading at the unknown point. Since time difference is equated to: [hours/24 x 360 = longitude in degrees], knowing the time at noon and the time at the prime meridian can be used to calculate longitude. Today, technological advancements with satellites and computer technologies have made it possible to measure longitude with a handheld global positioning system (GPS). Measurements taken with a GPS can be taken easily with levels of accuracy that would even amaze John Harrison, who dedicated 50 years of his life to find a solution to measure longitude.

Today, middle school science teachers have a unique opportunity to harness some of the same technologies that are available to scientists. For example, a graphing calculator can be linked to a probe (e.g., voltage probes, light sensors, and motion detectors) and data can be collected from laboratory activities and, with great precision, later uploaded to computer software. For life science lessons, a teacher can purchase a video microscope for about $300.00 and have students digitize still images of specimens. Most accessible, however, is the kaleidoscope of technological applications available for teachers via the internet. For example, using a directory of websites, teachers can connect their classrooms to images from a VolcanoCam focused on the crater of

Science as Inquiry Standards

In the vision presented by the Standards, inquiry is a step beyond "science as a process," in which students learn skills, such as observation, inference, and experimentation. The new vision includes the "processes of science" and requires that students combine processes and scientific knowledge as they use scientific reasoning and critical thinking to develop their understanding of science. Engaging students in inquiry helps students develop

- Understanding of scientific concepts.
- An appreciation of "how we know" what we know in science.
- Understanding of the nature of science.
- Skills necessary to become independent inquirers about the natural world.
- The dispositions to use the skills, abilities, and attitudes associated with science.

Science as inquiry is basic to science education and a controlling principle in the ultimate organization and selection of students' activities. The standards on inquiry highlight the ability to conduct inquiry and develop understanding about scientific inquiry. Students at all grade levels and in every domain of science should have the opportunity to use scientific inquiry and develop the ability to think and act in ways associated with inquiry, including asking questions, planning and conducting investigations, using appropriate tools and techniques to gather data, thinking critically and logically about relationships between evidence and explanations, constructing and analyzing alternative explanations, and communicating scientific arguments. Table 1 shows the standards for inquiry. The science as inquiry standards are described in terms of activities resulting in student development of certain abilities and in terms of student understanding of inquiry.

Science as Inquiry Standards

Levels K–4	Levels 5–6	Levels 9–12
Abilities necessary to do scientific inquiry	Abilities necessary to do scientific inquiry	Abilities necessary to do scientific inquiry
Understanding about scientific inquiry	Understanding about scientific inquiry	Understanding about scientific inquiry

Note: All of the material in this box was taken directly from the NRC Standards.

Science and Technology Standards

The science and technology standards in Table 2 present students with opportunities to develop decision-making abilities about the natural world that surrounds them. The science and technology standards stress that science is linked to technology as an ongoing process. For example, an engineer uses the fundamentals of science as building blocks to create useful products. The emphasis here, however, is that technology can be used as an essential tool for students to engage in the process of scientific inquiry.

Science and Technology Standards

Levels K–4	Levels 5–8	Levels 9–12
Abilities distinuish between natural objects and objects made by humans	Abilities of technological design	Abilities of technological design
Understanding about science and technology	Understanding about science and technology	Understanding about science technology

Therefore, as suggested in the NRC Standards, technology is "...a complement to the abilities developed in the science as inquiry standards, these standards call for students to develop abilities to identify and state a problem, design a solution—including a cost and risk-and-benefit analysis—implement a solution, and evaluate the solution. Science as inquiry is parallel to technology as design. Both standards emphasize student development of abilities and understanding."

Mount St. Helens, or to a schematic of a volcanic eruption that looks real and is more interesting to young students than the baking soda and vinegar model that has been used as a simulation in the past by teachers. If teachers are covering a unit on volcanoes, they can show students websites that describe the eruption of Mount Vesuvius in Italy in 79 AD and Krakatoa in the Indonesian arc in 1883. As long as inquiry-based lessons are in compliance with the tenets of the National Science Education Standards, teachers can use technology as a tool for inquiry in the framework outlined in the National Standards.

Conclusion

Technology-based inquiry is a pedagogical approach for middle school science teachers to give students the opportunity to use the latest tools to explore the natural world. Through hands-on experiences with the graphing calculators, calculator-based labs (CBLs), personal digital assistants (PDAs), global positioning systems (GPS), geographical information systems systems (GIS), and other emerging technologies, science teachers can develop and integrate the skills that can make inquiry-based learning meaningful for students. Hopefully, this book will show you how the tools of today can be implemented for technology-based inquiry in your middle school classroom.

References

Devore, P. W. 1980. *Technology: An introduction.* Worcester, MA: Davis Publications.

Kovich, F. J., and R. W. Shahan. 1980. Albert the Great: Commemorative essays. Norman, OK: University of Oklahoma Press.

Popper, K. 1972. Objective knowledge: An evolutionary approach. New York, NY: Oxford University Press.

National Research Council (NRC). 1996. *National science education standards.* Washington, DC: National Academy Press.

Trefil, J., and R. M. Hazen. 2004. *The sciences: An integrated approach.* Hoboken, NJ: John Wiley and Sons.

Time for Class

EDWIN P. CHRISTMANN

One of the most abstract concepts that you will teach to your students is the concept of time. Usually introduced at the beginning of the school year, the concept of time is taught along with measurements and scientific units such as length, mass, and volume (NRC 1996). However, unlike length, mass, and volume, time can be a very confusing concept to understand. Confusion about time is not restricted to students, but many others as well. For instance, when St. Augustine was asked, "What is time?" he responded, "If no one asks me, I know; if I want to explain it to a questioner, I do not know."

Time duration

Our 24-hour clock is based on the time it takes for Earth to make one complete rotation. Keep in mind that the Earth's circumference is about 40,073 km at the equator with an approximate rotational velocity of about 1,670 kph at the equator. (To find out the approximate rotational speed at your location, visit *http://pws.gamewood.net/~beaton/spinvb.htm.*) Recall, the equation for velocity is $v = d/t$, where, v = velocity, d = distance, and t = time, which can be used to determine the quantity of time in a day. To determine the number of hours in a day, all we have to do is rearrange the formula algebraically, so that $t = v/d$.

Subsequently, our equation results in a calculation of 24,900/1038 = 23.99 hours at the equator, which is very close to a 24-hour day. As you can see, however, our number did not come out to exactly 24 hours, which is the standard measure for one Earth day. In the course of a year, the Earth travels around the Sun in 365 days, 6 hours, 9 minutes, and 9.54 seconds. Our calendar, the Gregorian Calendar, has a correction built into it known as a leap year. Every four years we add an extra day, February 29, to the calendar as an adjustment. Although not a perfect system, it works. If this correction were not made, after 700 years we would have winter in July and summer in January.

Time standard

The universal unit of time is the second, which is 1/60 of a minute. A minute is 1/60 of an hour and an hour is 1/24 of a day. In 1967, to establish a more exact system of measurement, the Thirteenth General Conference on Weights and Measures defined the second as "the duration of 9,192,631,770 periods of the radiation of the cesium-133 atom." The result was the atomic clock, now operated by the National Institute of Standards and Technology (NIST) in cooperation with the U.S. Naval Observatory. The atomic clock and the hydrogen maser clocks operated by the U.S. Naval Observatory serve as the official source of time for the Department of Defense (DoD) and the Global Positioning System (GPS), and these clocks set the standard of time for the United States of America.

Your students can access the official time via the internet by visiting www.time.gov, which is operated by NIST and the U.S. Naval Observatory. If you do not have access to the internet, another option is to call the U.S. Naval Observatory at (202) 762-1401 to hear the exact time.

Classroom extensions

By having your students synchronize their watches to the exact time, they will be able to observe how a watch gains or loses time progressively. Even the finest certified chronometers have a daily margin of error between –4 and +6 seconds per day.

Based on the National Mathematics and Science Standards, a brief discussion on the history of early and modern chronometers, mechanical clocks, and quartz clocks should be of interest to your students. For a detailed history of time, visit *http://physics.nist.gov/GenInt/Time/time.html*, a site supported by the National Institute of Science and Technology.

Conclusion

The accurate measurement of time is crucial in all the sciences. Whether time is being used in an equation for calculations of velocity or to measure geological eras, it is the foundation for most scientific concepts. Therefore, by having students explore the technological advances that have improved the measurement of time, you will give them a better foundation for further scientific study.

Time resources on the web

A research guide for students
www.aresearchguide.com/time.html
It's about time
http://whyfiles.org/078time/index.php
Latitude and longitude
www.broward.org/library/bienes/lii14010.htm
U.S. Naval Observatory master clock
http://tycho.usno.navy.mil/cgi-bin/anim

References

National Council of Teachers of Mathematics. 2000. *Principles and standards for school mathematics.* Reston, VA: Author.

National Research Council (NRC). 1996. *National science education standards.* Washington, DC: National Academy Press.

Converting With Confidence

EDWIN P. CHRISTMANN

Around 1790, Thomas Jefferson petitioned Congress to adopt a system of dimensions and units based on multiples of 10. Fearing that such a system might impede trade with Britain, Congress rejected Jefferson's proposal and selected the English system of measurement, which originated in the 13th century. Consequently, our industrial revolution occurred during an era when force was measured in pounds, length in feet, and volume in pints and quarts.

In 1960, the General Conference on Weights and Measures recommended the International System of Units (SI) be adopted worldwide. In 1975, President Gerald Ford signed the Metric Conversion Act with the hope of converting the United States to the metric system. Unfortunately, we have made little progress toward this goal, thanks in part to the ubiquitous use of English measurements for everything from our speed limit signs to the temperature settings on our ovens. Thus, for the past 20 years American science teachers have been teaching students how to make conversions between the two systems (Weaver 1977).

The National Council of Teachers of Mathematics (NCTM) recommends that the math and science curricula for students in grades five through eight include explorations of measurement conversions based on different units (NCTM 1998). Moreover, the NCTM suggests,

> *In the United States, given the customary English system of measurement is still prevalent, students from elementary grades through high school should learn both customary and metric systems. Students should understand both systems, make conversions easily within systems, and estimate measurements fluently in both. As an example, a student might say, "I live a mile from the school. That is about two kilometers." Students will find it useful to know a few English-metric equivalents.*

Likewise, as outlined in the National Science Education Standards, middle school students should be able to "observe and measure characteristic properties, such as boiling and melting points, solubility, and simple chemical changes of pure substances" (NRC 1996). Once your students have demonstrated their proficiency in making conversions, you may want to let them use current technological tools to check their work. This will reduce the number of errors in their work, free up time for labs and other work, and allow them to focus on the science being taught instead of on long division.

Teaching conversions

During a middle school physical science unit, students are often required to take measurements, such as temperature readings. If the proper type of thermometer is not available, students must then convert their measurements from English units to SI units. For example, to

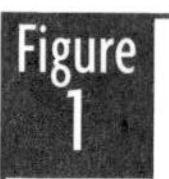

Conversion websites

Convert It:
www.image-ination.com/test_maker/convert.html

Entisoft Units:
www.entisoft.com/esunits2.htm

International French Property Metric Conversion Tables:
http://convert.french-property.co.uk

MegaConverter:
www.megaconverter.com

Science Made Simple Online Metric Converter:
http://www.sciencemadesimple.com/conversions.html

University of Berlin Conversion of Units:
www.chemie.fu-berlin.de/chemistry/general/units_en.html

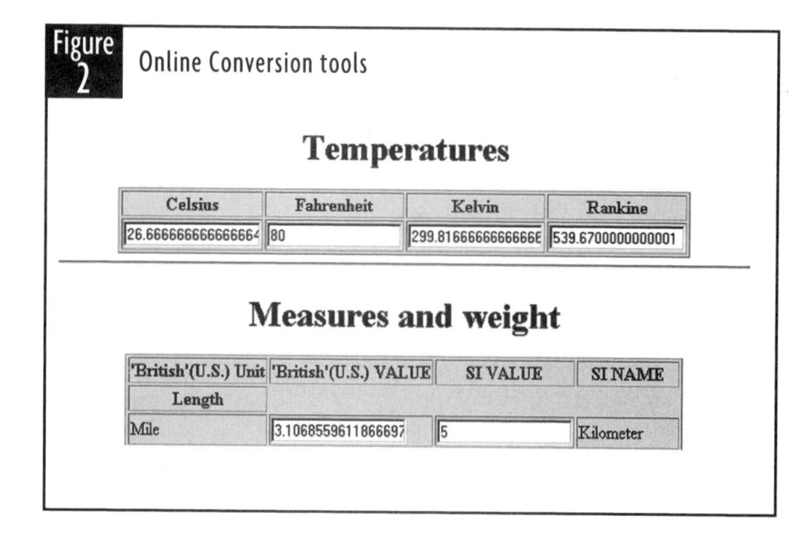

Temperatures

Celsius	Fahrenheit	Kelvin	Rankine
26.666666666666664	80	299.8166666666666	539.6700000000001

Measures and weight

'British'(U.S.) Unit	'British'(U.S.) VALUE	SI VALUE	SI NAME
Length			
Mile	3.1068559611866697	5	Kilometer

Figure 2 Online Conversion tools

convert a temperature reading, students would use the following formulas:

Fahrenheit to Celsius t°C = 5/9(t°F – 32°) or Celsius to Fahrenheit t°F = 9/5(t°C + 32°)

where t°C represents temperature in Celsius and t°F represents temperature in Fahrenheit.

Correspondingly, if a middle school teacher asked "If today's forecast is for a high of 80°F, what does that translate to in degrees Celsius?" The student would calculate the following:

t°C = 5/9(80°F – 32°)

t°C = 26.67°C

This relatively simple conversion can cut into valuable class time, especially if multiple conversions are needed. Over the course of a year, eliminating the need for such busy work could allow you to squeeze in an extra lab or two or to conduct extension activities that reinforce the lessons being taught.

To achieve this goal, students should check their work on one of several available conversion websites (see Figure 1). One of the simplest is Convert It, which allows students to simply fill in the value next to the English unit they want to convert and then hit the Tab key. The conversion appears in an adjacent box next to the appropriate SI unit. Students can convert from SI to English units just as easily.

Perhaps, the area that science teachers will find most useful is the measures-and-weights conversion boxes, which convert units of length, area, volume, capacity, mass, force, pressure, power, speed, and heat. For example, if a student wanted to know how many miles there are in a five-kilometer cross-country race, they would simply plug in the appropriate value in the kilometer box and hit the Tab key (see Figure 2).

Access denied?

For those of you who don't have access to the internet in the classroom, you have several options for simplifying the conversion process for your students. To start, several of the conversion websites allow you to download shareware versions of their conversion programs. This allows you to download the software at a different location and transfer it to your classroom computer.

If you don't have a computer in the classroom, you can always use calculators. Many calculators offer conversion features; for example, Texas Instruments' TI-73 graphing calculator is specifically designed for middle school mathematics and science and is an excellent tool for making English and metric conversions. If your students don't own their own calculators, keep a few on hand to loan out during class.

Summing it up

The applications for web-based conversion sites and the calculators are virtually endless for science teachers. I hope the ideas presented here will generate additional professional discussion on technology-based classroom conversion applications for middle school science teachers, as well as on how to better prepare students for the national and global competition of the twenty-first century.

References

National Council of Teachers of Mathematics (NCTM). 1998. NCTM principles and standards for school mathematics electronic version: Discussion draft. Reston, VA: Author.

National Research Council (NRC). 1996. National science education standards. Washington, DC: National Academy Press.

Weaver, K.F. 1977. How soon will we measure in metric? National Geographic 152 (August): 287–294.

A Different Phase Change

LYNDSAY B. LINK AND EDWIN P. CHRISTMANN

In physical science, we often use water to demonstrate change of state. However, this teacher demonstration will expose students to unfamiliar substances whose phase changes can be compared to those of water. Subsequently, a phase change that is in reverse of water's heating curve, cooling from liquid to solid, will show students that substances can freeze at room temperature. By observing the final graph, students will be able to observe that the terms *freezing point* and *melting point* are synonymous. Students will also become familiar with the latest technology-based tools that are now available for scientific inquiry.

What follows are the instructions for the teacher demonstration. Due to the hazardous nature of the chemicals, students must not be allowed to conduct this experiment. Only a trained teacher using a fume hood and wearing appropriate protective clothing and safety goggles can perform this demonstration.

Materials (for each group)

- glass test tube
- unknown substance (approximately 5 mL of stearic acid or lauric acid; both are solid at room temperature with a melting point less than 100°C)
- hotplate and GFCI protected electrical receptacle
- 2 medium-sized beakers
- water
- TI-73 graphing calculator (with DataMate software)
- TI Viewscreen (optional) for projecting calculator screen
- CBL2 and cradle
- temperature probe
- unit-to-unit cable
- calculator-to-computer cable
- computer with TI Connect software
- chemical splash goggles
- apron
- hand protection
- fume hood

Procedure

1. Fill half of a medium-sized beaker with water and place on a hot plate. Heat the water until boiling. While waiting for the water to heat, continue setting up the experiment.
2. Assemble the probe system as follows:
   a. Insert the upper end of the calculator into the cradle.
   b. Press down on the lower end of the calculator until it snaps into place.
   c. Slide the back of the cradle onto the front of the CBL2 until it clicks into place.
   d. Plug one end of the 6-inch unit-to-unit cable into the I/O port in the end of the CBL2, and plug the other end of the cable into the I/O port in the end of the calculator.
   e. Attach the temperature probe to channel one.

APPLICATIONS
1: FINANCE
2: AREA FORM
3: CBL/CBR
4: CABRI JR
5: CHEM BIO
6: CTLG HELP
7↓DATAMAP

3

DataMate is number 7 on this screen.

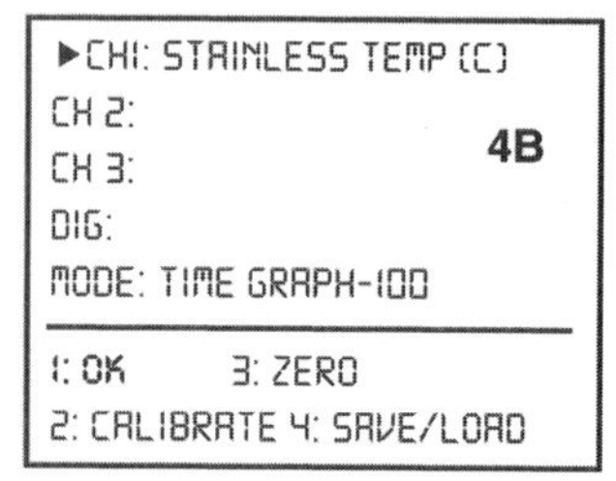

Press 1 here.

SELECT SENSOR
1: TEMPERATURE
2: PH
3: CONDUCTIVITY
4: PRESSURE
5: FORCE
6: HEART RATE
7: MORE
8: RETURN TO SETUP SCREEN

4C

Press 1 here.

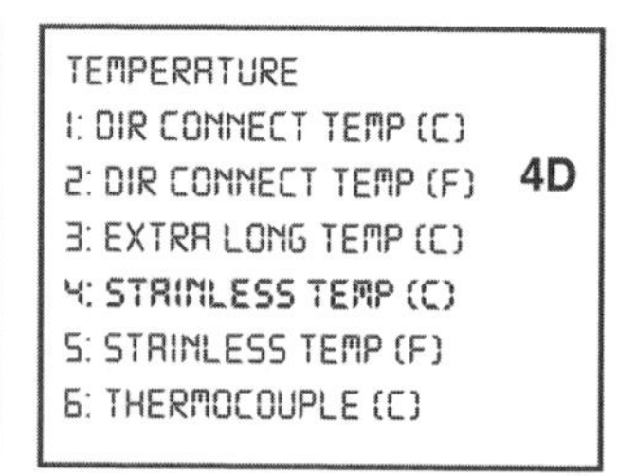

Press 4 here, then press 1.

3. Turn on the calculator. Press the PRGM button. You will see a screen like this: Press the number of the **DataMate** Application.
4. Set the temperature units to Celsius:
   a. Press **1** for **SETUP.**
   b. Press **ENTER** to choose the probe in Ch1.
   c. Press **1** for **TEMPERATURE.**
   d. Press **4** for **STAINLESS TEMP (C).**
   e. Press **1** for **OK.**
5. Set the experiment time to 20 minutes.
   a. Press **1** for **SETUP.**
   b. Move the cursor down to **MODE** and press **ENTER.**
   c. Press **2** for **TIME GRAPH.**
   d. Press **2** to **CHANGE TIME SETTINGS.**
   e. Type **5** as the **TIME BETWEEN SAMPLES IN SECONDS.**
   f. Type **240** as the **NUMBER OF SAMPLES.**
   g. Press **1** for **OK.**
   h. Press **1** for **OK.**
6. Obtain a sample of the substance to be identified. Place substance in a glass test tube. Heat the test tube by placing it in a water bath until melted and then continue heating one more minute.
7. Remove the test tube from the water bath and place in the second empty beaker. Place the temperature probe into the substance to be identified.
8. Press **2** to **START** collecting data. Data collection will take 20 minutes.
9. When data collection is complete, a graph of the data will appear.
10. When a substance is changing states (i.e., liquid to solid), there will be a small period of time where the temperature will actually remain constant. On a graph, this will appear as a section nearly horizontal.
11. Use the arrow keys on the calculator to move the cursor to the "change of state" or freezing point. The coordinates of this exact point should appear concurrently on the screen.
12. *Without changing the viewing screen*, attach the calculator to the computer. Open the TI Connect icon on the desktop. Choose TI Screen Capture. When the screen appears, copy and paste the image into a Word document.
13. Center and enlarge the graph and add a title and axis labels. Also include the following: student name(s), approximate freezing point, and unknown number (if more than one substance is used in the classroom). Print out this document.
14. Copies of this page should be distributed to students for analysis. As an alternative, you can hook your computer up to a projector to share the results onscreen at the front of the room.

Conclusion

Having students engage in technology-based inquiry activities is an excellent way for teachers to introduce topics that are driven by the National Science Education Standards (NRC 1996). Based on Content Standard B, this activity shows middle school students the different characteristic

CH 1: TEMP (C) 24.5
MODE: TIME GRAPH-180
1: SETUP 4: ANALYZE
2: START 5: TOOLS
3: GRAPH 6: QUIT
5A

Press 1 here.

CH1: STAINLESS TEMP (C)
CH 2:
CH 3:
DIG:
MODE: TIME GRAPH-100
1: OK 3: ZERO
2: CALIBRATE 4: SAVE/LOAD
5B

Move cursor to MODE and press enter here.

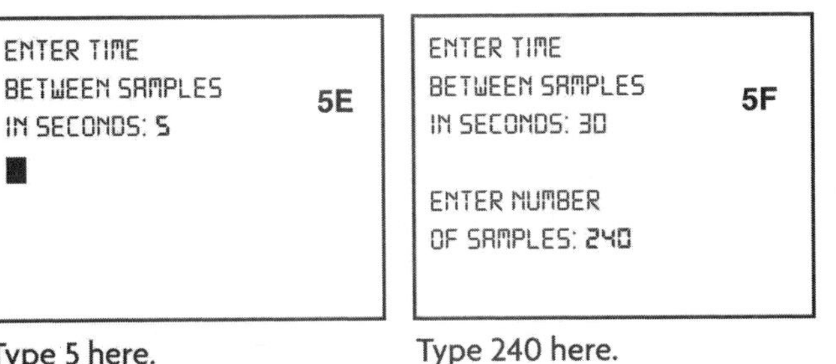

Type 5 here.

Type 240 here.

SELECT MODE
1: LOG DATA
2: TIME GRAPH
3: EVENTS WITH ENTRY
4: SINGLE POINT
5: SELECTED EVENTS
6: RETURN TO SETUP SCREEN
5C

Press 2 here.

TIME GRAPH SETTINGS
TIME INTERVAL: 1
NUMBER OF SAMPLES: 180
EXPERIMENT LENGTH: 180
1: OK 3: ADVANCED
2: CHANGE TIME SETTINGS
5D

Press 2 here.

CH 1: TEMP (C) 24.5
MODE: TIME GRAPH-180
1: SETUP 4: ANALYZE
2: START 5: TOOLS
3: GRAPH 6: QUIT
8

Press 2 here.

TEMP (C)
TIME(S)
X=265 Y=51.6667
9

The cursor marks the change of state (melting/freezing point).

properties of substances. Moreover, by using the TI-73 with a CBL2, students can take accurate measurements and have greater control over their experiments. In addition, the interactive graphs generated by the TI-73/CBL2 help students interpret the results of their experiments and can be used to create professional laboratory reports. Hopefully, this experiment opens the door for teachers to the endless possibilities for the coupling of a TI-73 with a CBL2 in the middle school science classroom.

Science Content Standards: 5-8

Science as Inquiry
CONTENT STANDARD A
As a result of activities in grades 5–8, all students should develop abilities necessary to do scientific inquiry.

- Use appropriate tools and techniques to gather, analyze, and interpret data.
- Communicate scientific procedures and explanations.
- Use mathematics in all aspects of scientific inquiry.

Physical Science
CONTENT STANDARD B
As a result of their activities in grades 5–8, all students should develop an understanding of

- properties and changes of properties in matter, and
- transfer of energy.

Science and Technology
CONTENT STANDARD E
As a result of activities in grades 5–8, all students should develop an understanding of

- science and technology.

Reference

National Research Council (NRC). 1996. *National science education standards.* Washington, DC: National Academy Press.

Resources

Texas Instruments Home Page
http://education.ti.com/educationportal
Download TI Connect Software
http://education.ti.com/us/product/accessory/

connectivity/down/download.html
Free CBL2 Tutorial
http://education.ti.com/us/training/online/freecbl2tutorials.html
Handbook of Chemistry and Physics
www.hbcpnetbase.com
ChemFinder
http://chemfinder.cambridgesoft.com
The MSDS Hyper Glossary: Freezing Point
www.ilpi.com/msds/ref/freezingpoint.html
General Chemistry—Liquids (Purdue University)
http://chemed.chem.purdue.edu/genchem/topicreview/bp/ch14/liquidsframe.html
Heating Curve (U. of Waterloo)
www.science.uwaterloo.ca/~cchieh/cact/c123/heating.html
CBL2 Getting Started Guidebook
http://education.ti.com/downloads/guidebooks/eng/cbl2-eng.pdf
Download TI Connect Software
http://education.ti.com/us/product/accessory/connectivity/down/download.html
Chemical Safety Database Searcher
http://ptcl.chem.ox.ac.uk/MSDS/msds-searcher.html

Temperature Tracking

JEFFREY LEHMAN AND EDWIN P. CHRISTMANN

In-depth internet

The authors of the National Science Education Standards envision a changing emphasis in school science that includes learning science in the context of inquiry, technology, and the history and nature of science (NRC 1996). Moreover, student management of ideas and information resulting from investigations over extended periods of time echo one of the major tenets of Project 2061: namely, do more by doing less (AAAS 1989). With an increasing number of schools acquiring internet access, middle school science teachers have an unprecedented opportunity to elaborate upon scientific concepts, while simultaneously helping students construct connections between the natural and the human-designed world.

The following example, which focuses on the concept of temperature, illustrates how in-depth study can meet multiple content standards for middle school science and mathematics that deal with measurement, the statistical analyses of data, changes in the properties of matter, the relationship between science and technology, and the nature and history of science (NCTM 1989).

Temperature tracking

Building upon students' elementary school experiences, middle school teachers often have students take temperature readings as part of a unit on measurement or within other science units, such as weather. Frequently, instruction about temperature is limited to a cursory discussion of the Fahrenheit and Celsius temperature scales, and possibly making conversions between the two scales. In part, this practice has resulted from teachers' perceived need to cover a breadth of science material.

Now, in addition to measuring and recording temperatures, students have access to a wealth of temperature data from various internet databases. By using such resources, science teachers are able to provide students with more in-depth instruction. For example, while studying weather-related factors for their local area, students can obtain

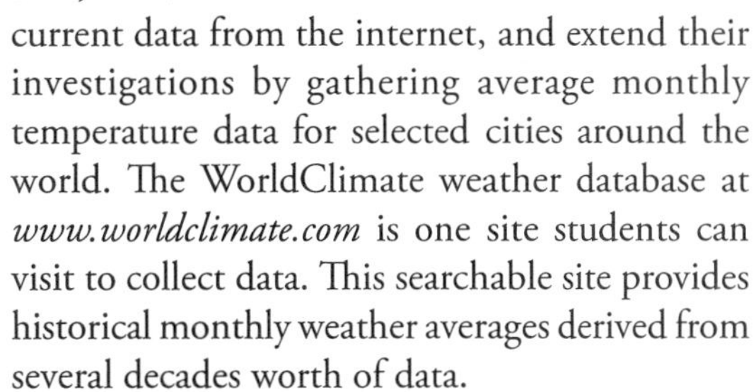

current data from the internet, and extend their investigations by gathering average monthly temperature data for selected cities around the world. The WorldClimate weather database at *www.worldclimate.com* is one site students can visit to collect data. This searchable site provides historical monthly weather averages derived from several decades worth of data.

Using these data, students can report for each city the mean yearly temperature, the minimum and maximum average monthly temperatures, the range of readings, and any noticeable patterns. For example, students can compare the range of average monthly temperatures for Seattle and Denver. Why is there less variation in Seattle?

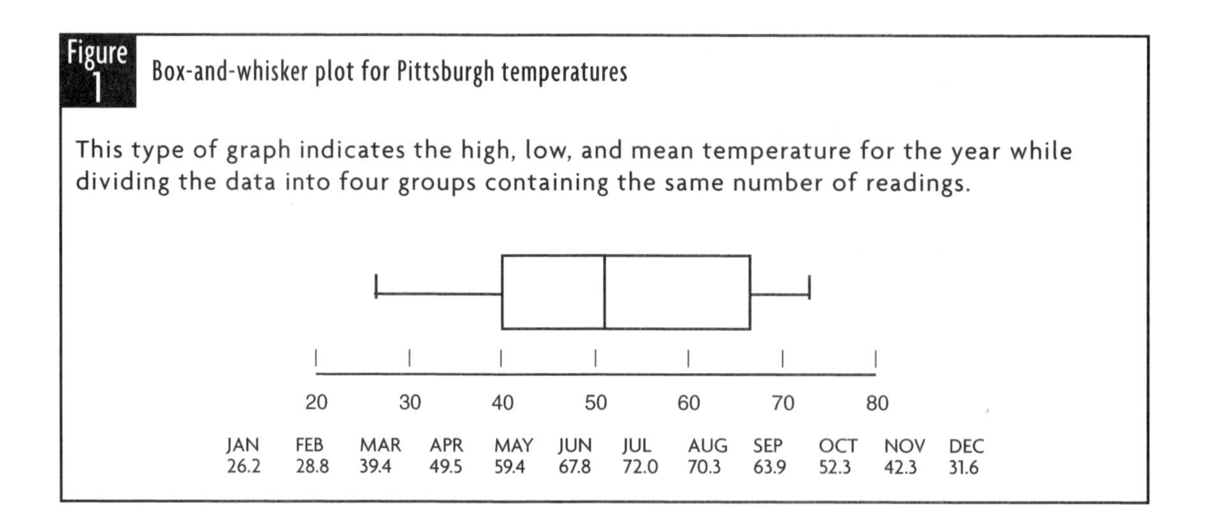

Figure 1 Box-and-whisker plot for Pittsburgh temperatures

This type of graph indicates the high, low, and mean temperature for the year while dividing the data into four groups containing the same number of readings.

How do other cities located near bodies of water compare with land-locked cities? Students can suggest possible explanations for these and other patterns they detect.

The data for a given city can also serve as the context for introducing students to a different type of graph—the box-and-whisker plot (see Figure 1). The 12 monthly average temperatures divide nicely into the four quartiles represented on the graph. Students can be asked to explain which seasons of the year are represented by the left whisker, the right whisker, and the two boxes. Again, they can be asked to compare plots from different cities around the world.

Finally, students can gather monthly temperature data for the current year to compare with historical data. A few minutes of recording time each month allows students to investigate a phenomenon over an extended period of time. Ongoing investigations of this type also give students the chance to make predictions and change conjectures as more data are collected—both desirable science objectives.

Several student- or teacher-generated questions prompt further student investigation. For example, students may want to know how long we've been keeping temperature records, who invented thermometers, or how many different temperature scales are used worldwide. Students can find answers to these and other questions from other internet sites, such as the About Temperatures site at *http://eo.ucar.edu/skymath/tmp2.html*. Both the cumulative nature of science and science as a human activity can be highlighted as students trace the development of early thermoscopes to sealed thermometers. The issues of calibration, using one or two fixed points in the development of scales, and altering the medium from colored water to alcohol and then to mercury can be explored. Moreover, small groups of students can assess the contributions of Galen, Galileo, Ferdinand II- Grand Duke of Tuscany, Robert Hook, Ole Roemer, Fahrenheit, and other scientists, and then report their findings to the entire class. Such a historical analysis provides teachers with the context to elaborate upon additional content topics, thus helping students make connections between temperature and the expansion/contraction of substances; the melting, freezing, and boiling points of materials; experimental error in measurements; the importance of the reliability (repeatability) of results; and the need for independent verification of results in science.

Additionally, students can search for original documents such as Anders Celsius's "Observations on Two Persistent Degrees on a Thermometer," a paper related to the origin of the Celsius scale, published in the *Annals of the Royal Swedish Academy of Science.* Not only does this endeavor introduce students to professional scientific literature, it also distinguishes between primary and secondary sources of information. Teachers can provide students with other primary documents throughout the year, as well as access to current scientific periodicals for alternative reading assignments.

The development of temperature scales and thermometers had implications beyond the study of temperature. Much of Fahrenheit's and Celsius's work occurred in the first half of the 18th century. What type of scientific inventions and discoveries came about just prior to and just following this work? Students can generate ideas on the importance of temperature in other areas of scientific advancement as they search a chronological listing of over 700 years of scientific discoveries. These can be found on sites such as Physics Chronology (*http://webplaza.pt.lu/fklaess/html/HISTORIA.HTML*).

Again, not only is the cumulative nature of science evident from such a listing, but students can hypothesize about how breakthroughs in temperature scale development paved the way for advancements in other science areas. Additional information about many scientists and their discoveries or inventions can be obtained from the archives at the California Institute of Technology (*http://archives.caltech.edu*). Once connected to this site, the user can access a searchable database of nearly 3,000 images related to the history of science, including pictures of scientific instruments.

The use of temperature probes interfaced with a microcomputer (Microcomputer-based laboratories—MBL) or a calculator (Calculator-based laboratories—CBL) can bring the discussion of temperature measurement into the present. Students should consider how probes function and why they look different from standard thermometers. In the lab, both probes and standard thermometers can be used and students can compare the results they obtain from the two devices. As a follow-up, students can search the internet to obtain information concerning temperature probes and their features (*www.vernier.com/probes/temp.html*).

In-depth analysis

The elaborated study described here for temperature illustrates the type of in-depth science instruction possible with many topics at the middle school level. The unprecedented availability of resources, such as those found on the internet, can facilitate the development of comprehensive curriculum materials. Recent findings from the *Trends in International Mathematics and Science Study* reveal that students in the United States are exposed to a larger number of science topics than students in other countries (*http://nces.ed.gov/timss*). However, our students seldom explore topics in any depth. Moreover, teachers continue to rely heavily on science textbooks that traditionally contain few, if any, examples of in-depth explorations of science topics. Hopefully, the ideas presented here will generate additional professional dialogue concerning the development of in-depth curriculum materials designed to propel our students into the 21st century.

References

American Association for the Advancement of Science (AAAS). 1989. *Project 2061: Science for all Americans*. Washington, DC: Author.

National Council of Teachers of Mathematics (NCTM). 1989. *Curriculum and evaluation standards for school mathematics*. Reston, VA: Author.

National Research Council (NRC). 1996. *National science education standards*. Washington, DC: National Academy Press.

Probing for Answers

EDWIN P. CHRISTMANN

Recent advances in data-collection technologies have given students access to the same types of probes used by scientists in research labs. Probes allow students to collect data on variables such as pH, acceleration, oxygen, light, and temperature. The devices and accompanying software are very user-friendly, allowing students to easily store, manipulate, and present data as charts, graphs, and tables. They are very similar to those used by scientists working in research laboratories across the United States.

However, before you invest in this technology, you should consider all of your options. Today's data-collection probes are designed to work with three major platforms: desktop and laptop computers, personal digital assistants (PDAs), and graphing calculators. Some probes are device-specific, but others will work across platforms with the use of adaptors.

The software that allows you to manipulate data collected with the probes is also an important consideration, especially if you prefer to export your data to spreadsheet or word processing programs to create lab reports and other documents. Make sure your probe software is compatible with the programs you want to use before you make your purchase.

You must also consider how you will be using the probes. If most of your labs are classroom based, then just about any probe will do. However, if you want to conduct data collection in the field, you should invest in probes that work with PDAs or graphing calculators.

Of course, the most important consideration for your school will be price. The most cost-effective approach is to buy probes that are compatible with hardware that you already have in your classroom. If you are starting from scratch, the least expensive system would be the graphing calculator/probe combination. The PDA/probe combination would be the next most affordable option, and the computer/probe combo would require the biggest startup investment.

Desktops and laptops

Many probeware packages are available for desktop and laptop systems. The packages vary in price depending on the number and accuracy of the probes, and the type of software included. One example is Pitsco's "Electronic Sensor Kits with Computer Interface" packages, which range in price from $59 to $79. Each package includes a complete sensor kit and software package. Another example is Onset Computer Corp's HOBO Data Logger system (see Figure 1). This kit is a bit more expensive at about $100, but they offer free middle school labs on their website (*www.iscienceproject.com/labs/6480_middleschool-labs.html*). Onset also reports accuracy levels for all of its sensors. For example, the thermister sensor has an accuracy level of +/- 0.2°C. Teachers

Figure 1

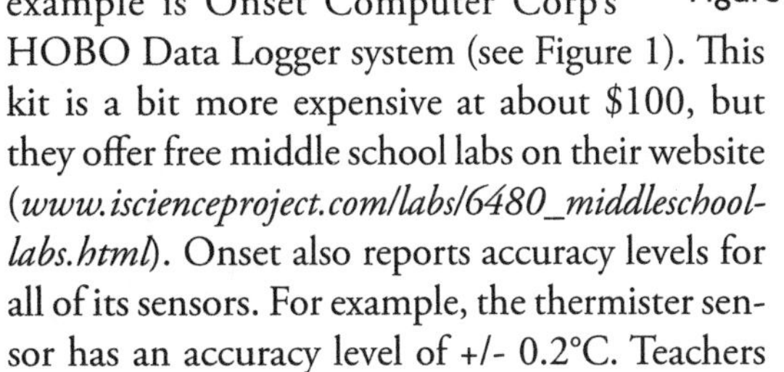

grades K–12 can also borrow a HOBO Data Logger for free to test it out. Visit *www.iscienceproject.com/contest/instantloaner.html* for details.

Personal digital assistants (PDAs)

Many probeware kits are also available for PDAs. For example, Pasco's Probeware Kit for Palm Handhelds ($229), includes probes to measure temperature, barometric pressure, humidity, and other variables. The kit also comes with a software package, Data Studio for Palm OS.

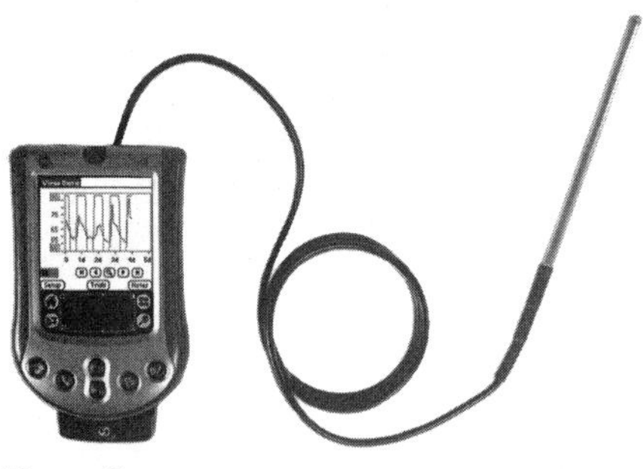

Figure 2

Another PDA software system that works very well is the ImagiProbe System ($249, see Figure 2). An advantage of the ImagiProbe is that it is designed to work with several brands of PDA (such as Palm and Visor). And, as long as your PDA has the proper adaptor, you can use it with probes made by different manufacturers. The ImagiProbe system also comes with a variety of lab activities so teachers can easily incorporate data-collection activities into course instruction.

Graphing calculators

CBL2 (about $200, Figure 3) from Texas Instruments is a cradle interface system that links a graphing calculator to probes. It is compatible with the TI-73, the graphing calculator recommended for the middle level. It is preprogrammed to work with all Vernier probes. Vernier offers a similar system, LabPro ($530, Figure 4), that has the added advantage of being able to work with PCs and PDAs. Probes included with this package include a motion detector, pH sensor, voltage probe, temperature probes, light sensor, exercise heart monitor, force sensor, conductivity probe, gas pressure sensor, and magnetic sensor.

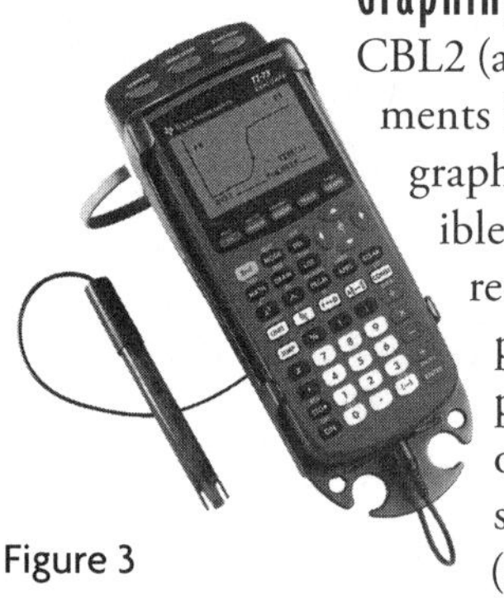

Figure 3

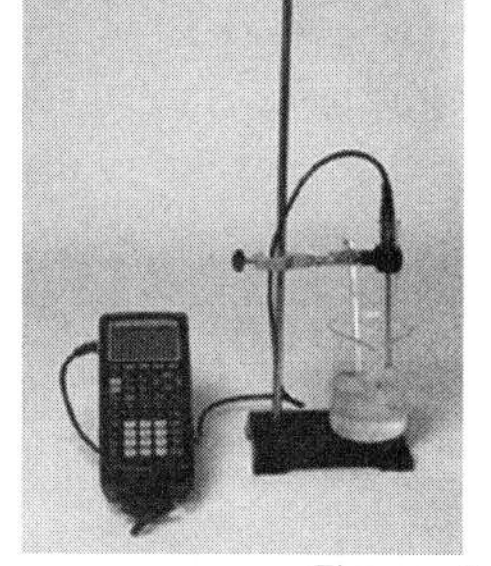

Figure 4

Conclusion

Obviously, for science teachers, the potential for data-collection systems is unlimited. However, if you choose to adopt the latest technologies, the most important consideration should be to select data-collection tools that promote basic understandings of scientific inquiry. It cannot be overemphasized that teachers should make purchasing decisions based on their annual budgets, needs, lab constraints, curriculum requirements, and the compatibility of the data-collection tool with the equipment that is currently in place. Hopefully, the ideas presented here will generate new and innovative ideas concerning the application of data collection tools in middle school science teaching and learning. After all, having your students probing for answers is what scientific inquiry is all about.

National Standards

The use of data-collection tools in the science classroom provides teachers and students with a unique opportunity to conduct inquiry-based activities that teach to the specifications of the NRC's (2000) report on scientific inquiry. Clearly, the use of laboratory activities to foster scientific inquiry with contemporary data-collection tools is an excellent way for science teachers to embrace the inquiry abilities that are proposed in the Content Standard for Science as Inquiry: Grades 5–8, which states:

- Identify questions that can be answered through scientific investigations.
- Design and conduct a scientific investigation.
- Use appropriate tools and techniques to gather, analyze, and interpret data.
- Develop descriptions, explanations, predictions, and models using evidence and explanations.

- Recognize and analyze alternative explanations and predictions.
- Communicate scientific procedures and explanations.
- Use mathematics in all aspects of scientific inquiry.

Online resources

Vernier:
www.vernier.com
Pitsco:
www.pitsco.com
Pasco:
www.pasco.com
HOBO Data Loggers:
www.iscienceproject.com
Texas Instruments:
http://education.ti.com/us/product/main.html

References

National Research Council (NRC). 1996. *National science education standards.* Washington, DC: National Academy Press.

National Research Council (NRC). 2000. *Inquiry and the national science education standards.* Washington, DC: National Academy Press.

Who Solved the Longitude Problem?

EDWIN P. CHRISTMANN

Few people would argue that classroom media applications are interesting, important, and occasionally irresistible. The omnipresence of media offers teachers many opportunities to incorporate attention-grabbing activities into classroom instruction, but teachers have few examples of how to effectively integrate media-based instruction into daily lesson plans. The following activity is a practical example of how a middle school science teacher can use a variety of media to conduct an in-depth exploration.

This activity requires a classroom set of Dava Sobel's book *Longitude: The True Story of a Lone Genius Who Solved the Greatest Scientific Problem of His Time*, a copy of A&E's drama "Longitude" (available on DVD or VHS), and internet access to view A&E's Classroom website (*www.aetv.com/class*). The objective of the activity is to teach students "about the difficulties of maritime exploration prior to invention of accurate clocks. They will learn about the scientific process that finally solved the problem of navigation and the political implications of naval dominance in the 18th century" (A&E 2002).

Historical overview

In 1492, Christopher Columbus's voyage to the "New World" bolstered the Age of Exploration. Unbelievably, Columbus found the "New World" without any reliable means of determining his longitude. His success was probably due to his mastery of latitude, which essentially allowed him to follow a straight path directly across the Atlantic.

Regretfully, however, not all explorers were as fortunate as Columbus. In response to the perils of sea travel, some of the greatest scientists and astronomers in history, including Galileo Galilei, Sir Isaac Newton, and Edmund Halley, searched for a solution to the longitude problem. However, all their efforts were unsuccessful. As a result, in 1714, the English Parliament offered a huge financial reward, known as The Longitude Prize, to any person who could successfully come up with a reliable system for calculating the longitude of a ship on the open sea. Finally, after 50 years of relentless pursuit, it was the clockmaker John Harrison who ultimately paved the way.

Mixing your media

Prior to assigning any book readings, teachers should pose a few discussion questions to get students thinking. The following two questions are based on samples found on A&E's classroom website, and would fit nicely into an Earth and Space Science unit:

Figure 1 A & E's classroom website

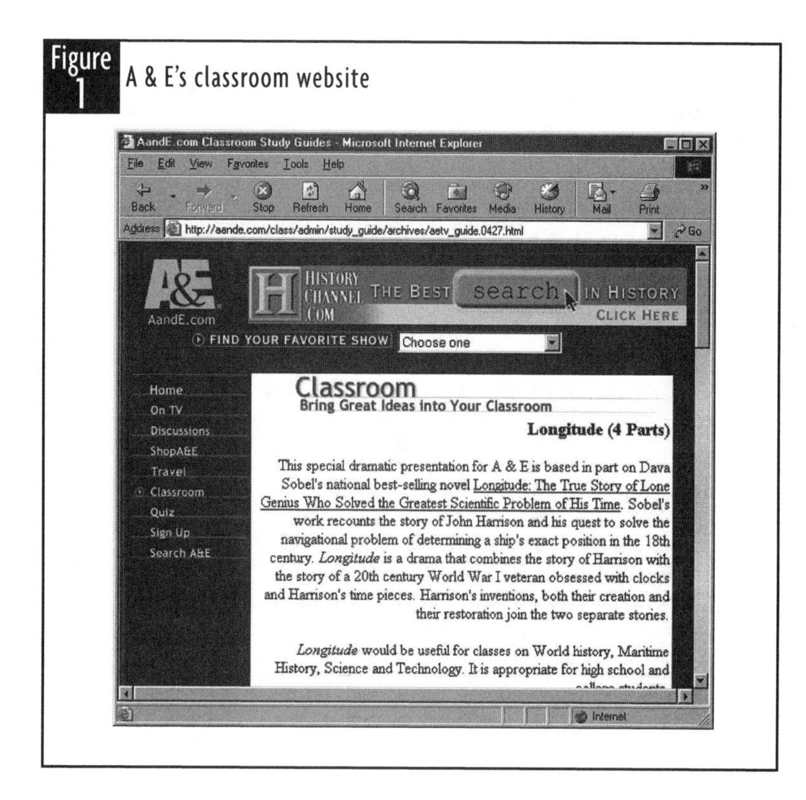

1. What is longitude? What is latitude? How are they used to indicate a person's position on the globe?

2. An hour's time difference between a ship and its starting point indicates how many degrees of longitude east or west?

After an initial introduction to the concept of longitude, students should begin reading *Longitude*. The website has a list of vocabulary words that students should review to help boost their reading comprehension. The words are linked to Merriam-Webster OnLine (*www.m-w.com*), to clarify definitions and pronunciations. These words include:

- allegation
- celestial
- centrifugal
- chronometer
- exemplary
- indulge
- mutiny
- obliged
- contrivance
- disembark
- patriotic
- squander

After giving students a week or so to read the novel (depending on the reading level of your students), assess them with an objective quiz that elicits critical thinking and includes follow-up essay questions. Again, teachers can use essay questions from the A&E Classroom website or create their own. The following are samples of two relevant, restricted-response, essay questions from A&E's Classroom website:

> There are two stories woven together in *Longitude*. Discuss how time is used to tell the story between the 18th century and the 20th century.
>
> Sailors in the 18th century often suffered from and died from scurvy. What is scurvy? How is it prevented?

Finally, after students have completed the book and quiz assignments, they now can watch A&E's drama "Longitude." As with any video, teachers will have to plan this activity well in advance so that it does not interfere with other important course content. Advantageously, however, this activity is well suited to be used as an interdisciplinary thematic unit. For example, science teachers can work with social studies teachers to combine this activity into a multidisciplinary unit. For example, a social studies teacher can explain how solving the longitude problem affected the British economy and maritime commerce.

As with the novel, A&E's Classroom website provides questions that can be used for testing and assessment after students have viewed the video. Moreover, as an extension activity, A&E also offers the following thought-provoking challenges that will make an excellent follow-up activity:

> Research modern methods of fixing navigational positions with those of the 18th century. Create a poster or chart illustrating these differences.

Write an essay comparing and contrasting Dava Sobel's book *Longitude: The True Story of a Lone Genius Who Solved the Greatest Scientific Problem of His Time* with A&E's presentation, "Longitude."

After reading the book and watching the video, students can find out more about John Harrison and longitude by searching websites such as *www.lincolnshire-web.co.uk/ lincolnshire-illustrious/john_harrison.htm*, which has a variety of pictures and historical articles that students should find interesting.

National Standards

This activity is aligned with Content Standard D: Earth in the Solar System, i.e., "Most objects in the solar system are in regular predictable motion" and Content Standard E: Science and Technology, i.e. "...students' work with scientific investigation can be complemented by activities that are meant to meet a human need, solve a human problem, or develop a product...." Moreover, teachers should develop lessons that emphasize inquiry, technology, and the history and nature of science for students in grades 5–8 (NRC 1996).

References

A&E. 2002. Longitude classroom lesson. Available online at *www.aetv.com/class/admin/study_guide/archives/aetv_guide.0427.html.* New York: A&E Television Networks.

National Research Council (NRC). 1996. *National science education standards.* Washington, DC: National Academy Press.

Sobel, D. 1995. *Longitude: The true story of a lone genius who solved the greatest scientific problem of his time.* New York: Penguin Books.

Celestial Observation Tools

EDWIN P. CHRISTMANN

For thousands of years, humans have been intrigued by the heavens. For example, some early civilizations, like the Maya in Mexico, were fascinated with the movement of the Sun across the sky. One of their celestial observation tools, which can still be found in the ancient Mexican city of Chichén Itzá, has a round astronomical observatory known as the Carocal *(www.mysteriousplaces.com/mayan/images2/MY028.JPG)*. This building, circa 250–900 AD, has several windows that point toward the equinox sunset. Amazingly, the position of each window directs an observer to the southernmost and northernmost points on the horizon where Venus rises.

Although the Carocal was useful for mapping the sky with the naked eye, it wasn't until 1609 that Galileo Galilei (1564–1642) built a powerful enough telescope to make detailed observations of celestial objects. With his telescope, Galileo was able to see the movement of the four moons around the equator of Jupiter (Figure 1). In all likelihood, this observation congealed Copernicus's heliocentric system theory (Benson 2002). At the time, however, heliocentric theory conflicted with the theological teachings of the Roman Catholic Church, which made Galileo a very controversial figure.

Fortunately, today's students don't have to erect specialized buildings or construct their own equipment to predict the location of and observe specific celestial bodies. With the click of a few buttons, modern technology can pinpoint the location of thousands of mapped objects in the universe. Once you know where to look, all you need is an inexpensive telescope to begin your exploration of the universe. The following activity, which asks students to estimate the orbits of the four innermost moons of Jupiter through direct observation, represents a practical example of how a middle school teacher can use technology to explore celestial objects.

To duplicate this experiment, students will need to have access to a relatively low-powered telescope. My recommendation is a refractor telescope with a 50mm objective lens, 150-power magnification, and an adjustable tripod mount. A telescope with these specifications usually costs less than $100. In addition, to locate the position of Jupiter, students will need a compass and access to the internet.

Finding Jupiter

Although Moon and space maps are readily available, several websites also provide a fast, interactive, and fun way to find celestial objects. In my opinion, the best website for this activity is *Sky & Telescope*'s Interactive Sky Chart. (See Figure 2 for web addresses and a list of software mentioned in this article.) Java Jupiter is another good site that allows you to determine the position of the moons as well as Jupiter's red spot for any given day and time. If you have a Personal Digital Assistant, such as the Palm Pilot, stargazing software is also available at a minimal expense for portable use (see page 99 for a description of this software).

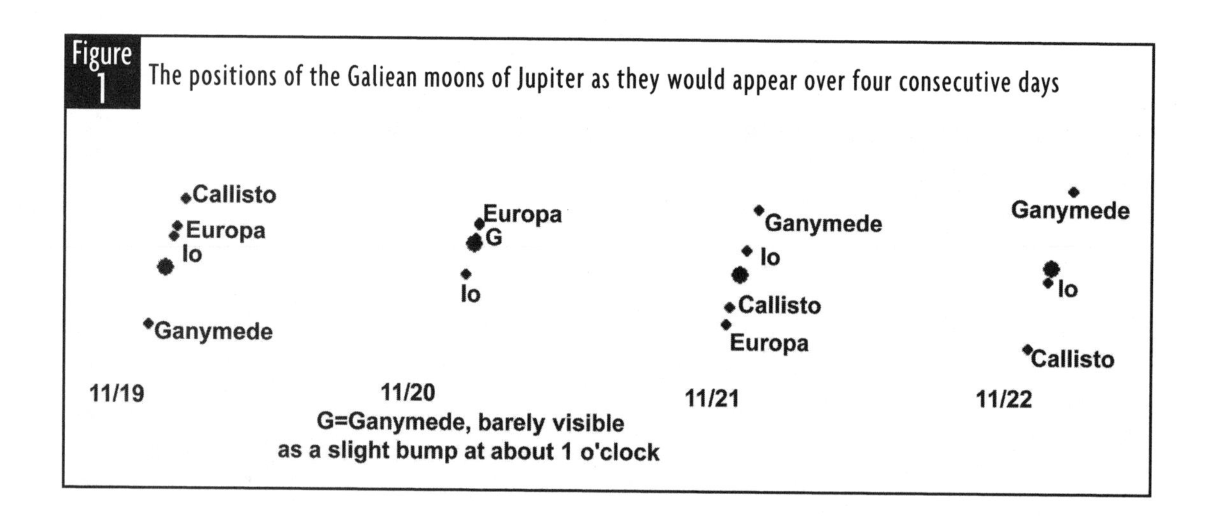

Figure 1 The positions of the Galiean moons of Jupiter as they would appear over four consecutive days

Figure 2 Internet links for celestial observation tools

Website	URL	Price
Sky Chart	http://skyandtelescope.com/observing/skychart/#	Free
PC-Sky	http://www.skyhound.com/pcsky.html	$39.95
Star Chart	www.telescope.com/content/inthesky/content6main.jsp?chart=true	Free
Java Jupiter	www.shallowsky.com/jupiter.html	Free

To find Jupiter using *Sky & Telescope's* Interactive Sky Chart, begin by entering your zip code. This configures the online chart to your latitude and longitude. Next, by rotating the All-Sky Chart's view of the night sky with your mouse, you can find Jupiter in the Selected View box. As an example, based on the Interactive Sky Chart that I used at the time of writing, at 9:00 p.m. on June 10, 2002, I found Jupiter in the WNW sky at about 8 degrees (see Figure 3). Remember, the position of Jupiter in the night sky and its position relative to other celestial objects will vary throughout the year. The best way to demonstrate how to find a celestial object using this website would be to project the site on a screen using a projector. However, you can also have students gather around the monitor in small groups if a projector is not available.

For this activity, students will need to know where to look in the night sky to find Jupiter. If they have internet access at home, they can visit the website for guidance immediately before making their observations. Students without access can be given a printout of the website to be used as a map that evening. Remember to adjust the view on the website to reflect the time at which the student will be making the observation that evening. All students will need to know how to use a compass and their maps to successfully locate Jupiter through a telescope.

Depending on your class size, a pair or team of students will be responsible for making observations on one of the 17 consecutive nights. Student pairings should be made based on their proximity to one another and access to transportation at night. Before a group takes home the telescope, they should receive training on how to care for and operate it. If you'd prefer not to loan out the telescope, you could also arrange to meet the groups at the school each night for a more supervised viewing. This would also allow you to help students locate the moons and create an accurate sketch of their position.

Background information

"Galileo did not invent the telescope (Dutch spectacle makers receive that credit), but he was the first to use the telescope to study the heavens systematically. His little telescope was poorer than even a cheap modern amateur telescope, but what he observed in the heavens rocked the very foundations of Aristotle's universe and the theological-philosophical worldview that it supported. It is said that what Galileo saw was so disturbing for some officials of the Church that they refused to even look through his telescope; they reasoned that the Devil was capable of making anything appear in the telescope, so it was best not to look through it." Moreover, "Galileo observed 4 points of light that changed their positions with time around the planet Jupiter. He concluded that these were objects in orbit around Jupiter. Indeed, they were the four brightest moons of Jupiter, which are now commonly called the Galilean moons. Galileo himself called them the Medicea Siderea, the Medician Stars" (From University of Tennessee's Online Journey through Astronomy website: *http://csep10.phys.utk.edu/ astr161/lect/ history/ galileo.html*).

Jupiter has 39 moons. Its four largest moons, the Galilean moons, are Ganymede, Callisto, Io, and Europa. For more information about Jupiter, visit *www.jpl.nasa.gov/ solar_system/planets/jupiter_index.html* and *www.jpl.nasa.gov/galileo.*

Figure 3 Sky & Telescope website screenshot

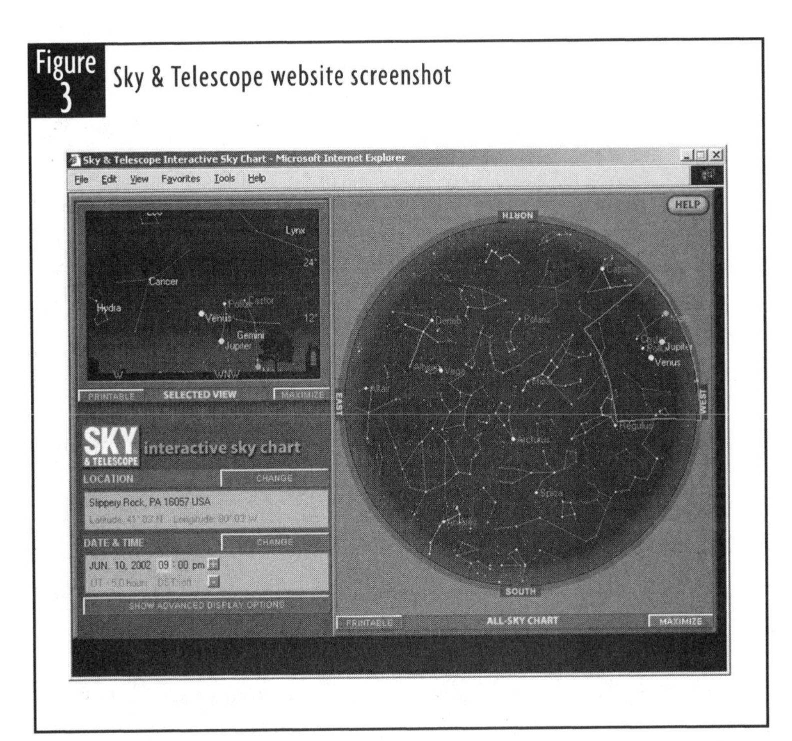

Moons of Jupiter

Objectives

You will

- observe the position of an object in the sky by describing its location relative to another celestial object,
- describe an object's motion by tracing and measuring its position over time,
- explain locations and movements that can be observed, and
- illustrate patterns of movement.

Equipment

- sketch paper
- access to the internet
- telescope
- compass

Procedure

At school or home, locate the position of Jupiter in the night sky for your location using the website or sky maps provided by your teacher.

Use a compass and telescope to find Jupiter in the night sky and note the position of its four innermost moons.

Create a sketch similar to Figure 3 showing the position and direction of travel of the four innermost moons relative to Jupiter. (Note: You may need to make several observations during the night to determine if the moon is moving left-to-right or right-to-left. Also, it is possible that a moon is behind the planet, so all four may not be visible each night.) Identify each of the moons and label them.

At the end of 17 days, your teacher will distribute copies of the sequential sketches made by all the groups. After you have studied the sketches, answer the following question.

Question

How many days does it take each moon to complete an orbit around Jupiter? Fill in the chart below. To approximate an orbit, you'll need to note the distance each moon appears from Jupiter and the direction it is traveling (left-to-right or right-to-left) and then refer to your sketches to track each moon as it circles the planet and returns to that position and direction of travel. It will be easier to estimate the outer moons' orbits, because they have longer orbits, but do your best to estimate an orbit for each moon.

Moon	Length of orbit in days
Io	
Europa	
Ganymede	
Callisto	

References

Benson, M. 2002. Centerpiece: A space in time. *Atlantic Monthly* 290(1): 91–112.

National Research Council (NRC). 1996. *National science education standards.* Washington, DC: National Academy Press.

National Standards

This activity fulfills Content Standard A: Science as Inquiry, Content Standard F: Earth in the Solar System, and Content Standard G: History and Nature of Science of the National Science Education Standards (NRC 1996).

Ready to Navigate: Classroom GPS Applications

ROBERT A. LUCKING AND EDWIN P. CHRISTMANN

Want to increase your students' awareness of their surroundings and broaden their understanding of their place in the world? Then start exploring with the help of a GPS device. As any hiker, boater, or new-car owner already knows, GPS is short for Global Positioning System, a satellite-based system that tells anyone, anywhere on Earth, at any time, exactly where they are. A total of 24 GPS satellites are located in orbits 11,000 nautical miles above the Earth. These satellites are monitored and coordinated by six ground-based facilities. With the help of a handheld device costing as little as $100, a person can receive data from the satellites to determine their exact location, distance from another location, altitude, and even speed if they happen to be moving. This technology allows students to explore directly the geographical dimensions of their world.

When you turn on your inexpensive and quite portable GPS device (available in watch form), it receives data at the speed of light from at least six of the GPS satellites in orbit. An internal clock on the satellite, that is accurate to within three nanoseconds, measures the time it takes for a signal to travel from the device to the satellite. Then the distance between the device and satellite is calculated by multiplying the time by the speed of light. It is then possible to determine the precise latitude, longitude, and altitude of the device by processing data from four separate satellites.

The potential contribution of GPS and related GIS (Geographic Information Systems) technology to education was observed shortly after the launching of the first geo-signaling satellites (Tinker 1992). However, only in the last few years have significant efforts gone into developing a wide array of rich teaching materials. Today, numerous websites dedicated to GPS-based education are available, which demonstrate teachers' growing interest in this technology and students' growing enthusiasm for GPS activities.

Students enthusiastically leap to exercises that require them to use a GPS receiver, an object no larger than a cell phone (see Figure 1). With it, students can capture remarkable details about their physical world. Before you actually put a GPS unit in students' hands, you will need to review the concepts of latitude and longitude. Specifically, students will need to know that measurements are taken in degrees, minutes, and seconds, and that one minute of latitude equals one nautical mile. (See page 27 for more information on longitude). One way of reinforcing the concepts is to display a nautical chart in the room and ask students to provide the longitude and latitude of several points at sea and along the

Figure 1 GPS receiver

The Garmin etrex Legent is one of the many GPS devices on the market.

Figure 2 GPS applications for PDAs

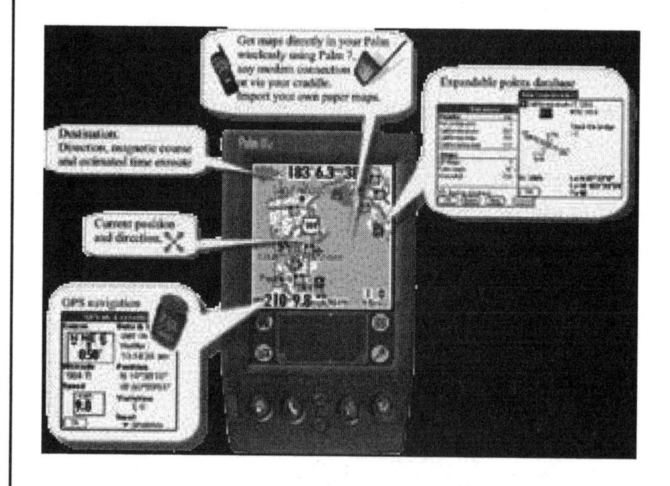

This image reflects software designed by GPS Pilot *(GPSpilot.com)*. GPS Pilot products offer everything from digital atlas functionality (with tracking and routing information) to GPS navigation assistance.

coast. You can also ask them to develop a route from Point A to Point B using several waypoints, and have them mark the latitude and longitude of each. Because navigation requires maintaining compass headings, ask them to determine the degree of angle for each of the turns they will be making at each waypoint along the route, and have them calculate the reciprocal angles of turn for the trip home. For example, if the heading from your school building to the school maintenance building is 140 degrees, then students must add 180 degrees to that sum to know that the return heading is 320 degrees. You can also provide students with the latitude and longitude of two specific locations on the chart and have them determine the distance between the two.

But the fun begins when you actually give your students a GPS receiver. Even the most inexpensive units provide the user's present latitude and longitude, average velocity, and calculate the bearing of a distant point. Using perform these capabilities, students can perform exercises that ask them to determine distances between the school and landmarks in their hometown, across the country, or around the globe. Students can then compute how far from a given city they may be and how long it would take to get there on foot, in a car, or on a plane. In addition, you could arrange a landmark scavenger hunt in a local park, giving the latitude and longitude for 10 different landmarks and having students use the GPS device to track them down. As they traveled from landmark to landmark, students would need to record the distances traveled, degrees of turn, and travel times. To integrate a history lesson, you can have students research the history of early naval exploration, including the development of the sextant, and fundamental concepts involving longitude and latitude. For a geography extension, have students determine the distance between the true North Pole and the magnetic north pole.

All of these possibilities are amplified when you combine the receiver technology with that of a PDA (see Figure 2). With a cable connecting a

Figure 3 GPS on hand

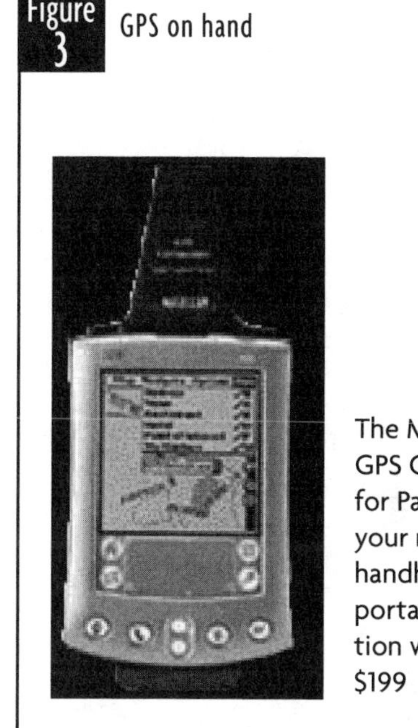

The Magellan GPS Companion for Palm® turns your m500 Series handheld into a portable navigation wizard. Price: $199

Figure 4 PalmWorld

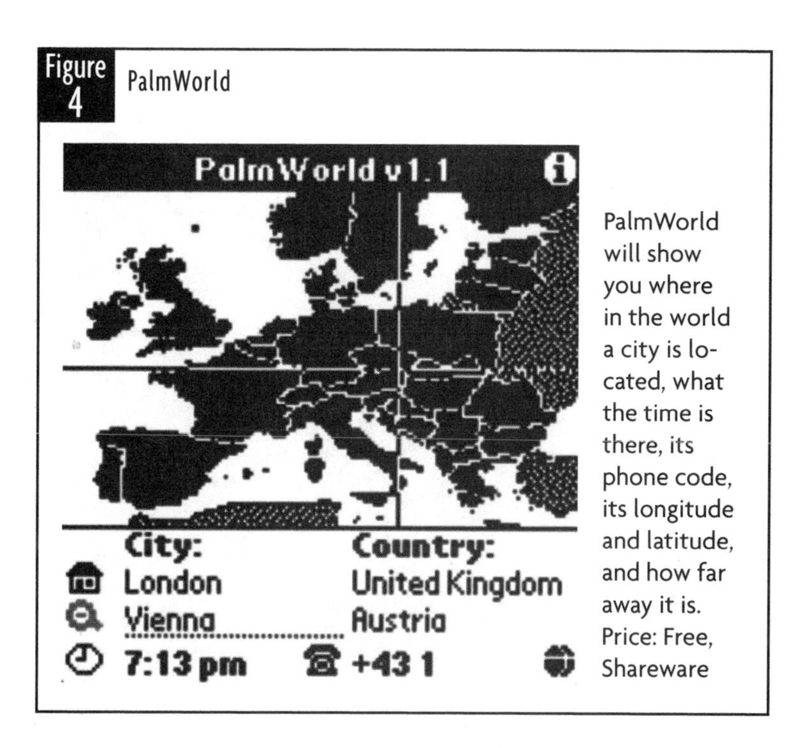

PalmWorld will show you where in the world a city is located, what the time is there, its phone code, its longitude and latitude, and how far away it is. Price: Free, Shareware

GPS receiver to your handheld computer, PDA software for under $200 now allows you to watch your position on a map or chart that appears on the miniature screen in exactly the same way that very expensive ship and airplane navigation units work (see Figure 3). GPS software for the PDA is available at *www.gpspilot.com/products/index.shtml.* Using this software, you can set your students on a geographical adventure around your school building or school grounds.

Another great resource for teachers can be found at the NASA website entitled From a Distance: An Introduction to Remote Sensing/GIS/GPS (*www.education.ssc.nasa.gov/fad/default.asp*). It includes many lesson plans, for different age groups, that are tied to the National Science Education Standards.

A piece of shareware called PalmWorld (Figure 4) will show you where in the world a city is located, the local time, its phone code, its longitude and latitude, and the distance from your location. Using this software, you can ask students to create a map showing the major cities around the country or globe, and indicate the distances between them. This will improve their global understanding and geographic awareness. For example, when students realize that it is 2,335 miles from Washington, DC to San Diego, they can predict that it is nearly the same number of miles from Washington to Los Angeles and a little more than half that distance to Denver. In completing these exercises, students are formulating spatial relationships derived from their data collection, plotting, and manipulation.

Conclusion

Obviously, the potential for enriched science instruction using GPS is virtually unlimited (see below for a description of other useful PDA software programs). Additional professional discussion concerning GPS applications for middle school science teachers is merited, and the focus of that conversation must be the preparation of students for the national and global dynamics of the 21st century.

Figure 5 GPS Software for PDA

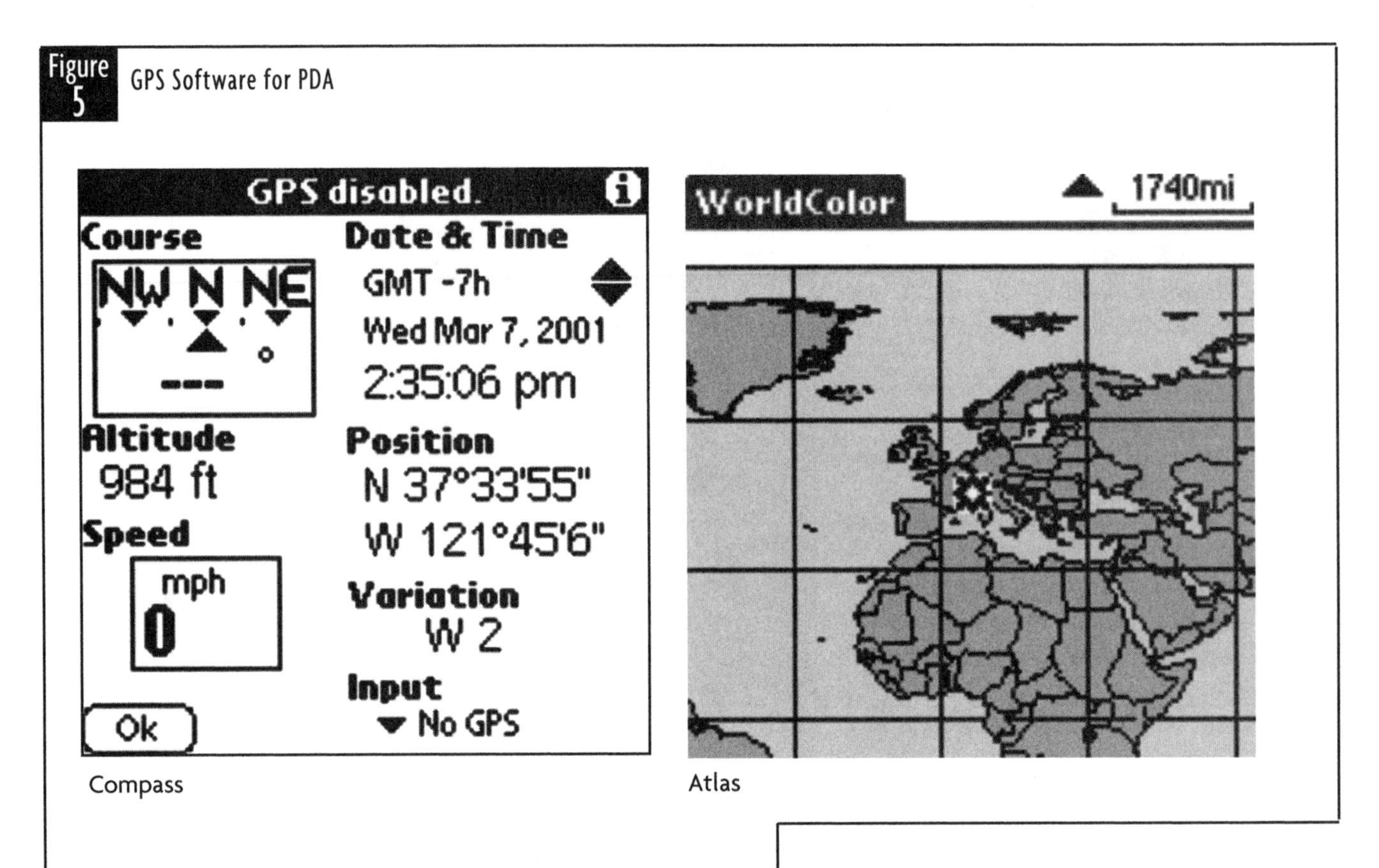

Compass

Atlas

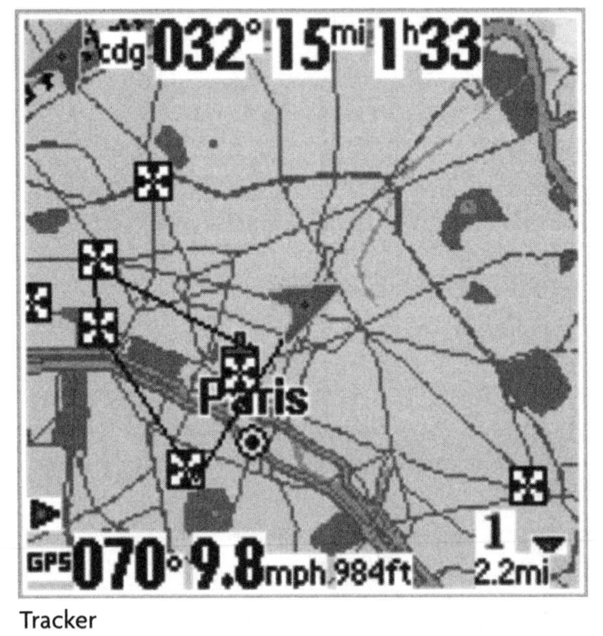

Tracker

GPS software for PDAs

Compass

Compass is a GPS application for the Palm. The main screen provides you with your bearing, current speed, current latitude and longitude, altitude, magnetic variation. In addition, you get the functionality of a built-in atomic clock because the program shows you the date and time of the satellites (in universal time). The number of satellites in view and the validity of their signal are shown in the title bar. You can easily change the units of measure. Price: $24.95

Atlas

Atlas is a GPS application for the Palm. You add a map appropriate for your operating system (including color if your Palm supports it), plug in your GPS receiver, and Atlas co-pilots you to your destination by pinpointing your position and scrolling the map in real time. It also allows you to add your own reference points. You can import maps from the internet or simply scan them on your home computer. Calibration of new maps is simple. You can easily change the units of measure, zoom in and out, or even beam your map to a friend's PDA. Price: $39.95

Tracker

Tracker lets you display black-and-white and color (if your PDA supports it) maps for traveling. When connected to a GPS receiver it can display your current position and movements and also record your route. It offers easy scrolling and zooming. The Preferences menu lets you select the units in which the distance, speed, and altitude will be displayed, and you can also choose the types of information (landmarks, cities, airports) that will be displayed. Location information can be displayed as an icon or as a text marker. If a GPS receiver is connected, the title bar displays GPS navigational information such as current heading, speed, and altitude. Price: $49.95

Additional GPS software can be downloaded from *www.handango.com.* Many shareware programs are available, as are free trials of commercial products.

National Standards

GPS classroom applications are aligned with Content Standard D: Earth in the Solar System, i.e., "Most objects in the solar system are in regular predictable motion" and Content Standard E: Science and Technology, i.e., "...students' work with scientific investigation can be complemented by activities that are meant to meet a human need, solve a human problem, or develop a product...." for students in grades 5–8 (NRC 1996).

The integration of GPS into science instruction conforms to the tenets of the *National Education Technology Standards for Students* (ISTE 1998) as follows:

1. Basic operations and concepts
   - Students demonstrate a sound understanding of the nature and operation of technology systems.
   - Students are proficient in the use of technology.
2. Social, ethical, and human issues
   - Students understand the ethical, cultural, and societal issues related to technology.
   - Students practice responsible use of technology systems, information, and software.
   - Students develop positive attitudes toward technology uses that support lifelong learning, collaboration, personal pursuits, and productivity.
3. Technology productivity tools
   - Students use technology tools to enhance learning, increase productivity, and promote creativity.
   - Students use productivity tools to collaborate in constructing technology-enhanced models, prepare publications, and produce other creative works.
4. Technology communications tools
   - Students use telecommunications to collaborate, publish, and interact with peers, experts, and other audiences.
   - Students use a variety of media and formats to communicate information and ideas effectively to multiple audiences.
5. Technology research tools
   - Students use technology to locate, evaluate, and collect information from a variety of sources.
   - Students use technology tools to process data and report results.
   - Students evaluate and select new information resources and technological innovations based on the appropriateness for specific tasks.
6. Technology problem-solving and decision-making tools
   - Students use technology resources for

solving problems and making informed decisions.
- Students employ technology in the development of strategies for solving problems in the real world.

Internet resources

NASA's From a Distance: An Introduction to Remote Sensing/GIS/GPS:
www.education.ssc.nasa.gov/ltp

Harvard University's Using GPS in the Classroom:
cfa-www.harvard.edu/space_geodesy/ATLAS/classroom.html

GPS Palm Use:
www.eduscapes.com/tap/topic78.htm

GPS Primer:
http://www.aero.org/education/primers/gps

Satellite Navigation:
http://gps.faa.gov

References

International Society for Technology in Education (ISTE). 1998. *National educational technology standards for students.* Eugene, OR: ISTE Press.

National Research Council (NRC). 1996. *National science education standards.* Washington, DC: National Academy Press.

Tinker, R.F. 1992. Mapware: Educational applications of geographic information systems. *Journal of Science Education and Technology* 1(1): 35–48.

Layers of Information: Geographic Information Systems (GIS)

ROBERT A. LUCKING AND EDWIN P. CHRISTMANN

While many science educators work effectively within the boundaries of an instructional world driven by state assessments and a narrowly-prescribed curriculum, a smaller minority are quietly developing new teaching techniques that may well establish the way science is taught in the future. These teachers are experimenting with instructional approaches based on powerful technologies and interdisciplinary perspectives that hold a great deal of promise for placing students in active roles of exploring their physical universe in a "new geography." Moreover, while these efforts may at first appear to be on the periphery of science teaching, they may allow for the enrichment of the many topics of instruction present in well-conceived curricula. The core of these efforts center around GIS (Geographic Information Systems). Simply put, GIS combines layers of information to give you a better understanding of place (Figure 1). Which layers of information you combine depend on your purpose—finding the best location for a new store, analyzing environmental damage, viewing similar crimes in a city to detect a pattern, and so on. The layers are manipulated through a combination of computer hardware and software.

Related technologies such as desktop mapping, remote sensing, and database management systems are also being exploited as powerful teaching and learning tools. Collectively these technologies present opportunities for interdisciplinary teaching in which teachers hand over to students the tools to analyze data in visual forms never before possible, thus developing a richer understanding of geography, Earth science, physics, and biology.

Many teachers begin with a variety of new mapping options, which GIS can handle with aplomb, but this new form of technology can also handle far more complex tasks. GIS can store, manipulate, and project geographical data in very sophisticated ways. This software can not only display data such as those provided by the Census Bureau, but it can layer this information over various representations of land features. Additionally, it can display data in three-dimensional

Figure 1 Sample Layers

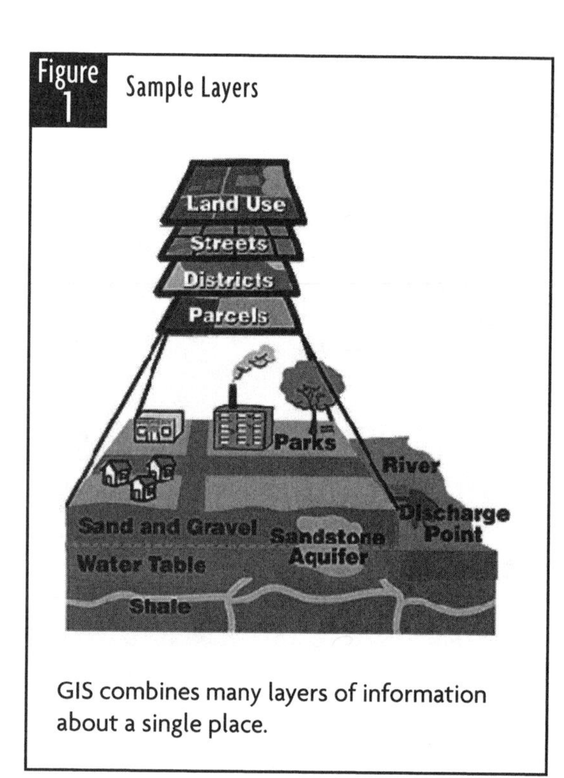

GIS combines many layers of information about a single place.

perspectives and depict data over time periods. By manipulating these layers of information, students can explore complex relationships in meaningful scientific inquiry. Given these qualities, the software lends itself to student analysis and places emphasis on observation, testing, and exploration.

The Environmental Systems Research Institute, Inc. (ESRI) has developed a GIS package called ArcView that combines mapping, database, and spreadsheet capability with a geographic representation engine. Other GIS software packages are also available, but most are prohibitively expensive for classroom use. However, ESRI has made one program, ArcVoyager, available specifically for K–12 education, and it is a very useful program (Audet and Ludwig 2000, Davis 2000). Additionally, ESRI has established a website (*www.esri.com/k-12*) specifically for teachers, where lesson plans and very useful sample data can be found.

The raw material of GIS exploration is digitized information about our Earth that holds the key to spatial patterns and relationships. These spatial data represent information about the location of objects found on Earth and their relationships, and it is stored in a database that can be queried in hundreds of ways. Therefore, GIS can depict a map of considerable detail, index it by latitude/longitude, and display a wealth of "attribute data" about the events or circumstances of that place. For example, GIS can store maps showing streets, rivers, and towns, but it can also show traffic patterns, water quality, and incidents of disease outbreaks.

The key to the power of inquiry of GIS rests in "shapefiles," views, and themes. In ArcView, a shapefile consists of three separate types of files relative to a map; it holds the points, lines, and polygons represented on the map, but it also includes files that index many kinds of detailed information relative to that map. ArcView then uses what are called *themes* to depict information, and themes are then arranged in layers. One layer might show all bodies of water, another might be county boundaries, and another city locations, and these can be clicked on and off with a mouse. The GIS software then responds to queries by depicting data graphically. With a couple of keystrokes, students can depict population density, and research real-world problems such as incidents of traffic congestion, and water quality violations.

GIS in schools

Many GIS projects are underway in schools across the nation. In Virginia, students are studying the spread of the West Nile virus. In Minnesota, students track wolves across the northern landscape. In Massachusetts, high school students research a simulated spill of toxic chemicals and manage the mock evacuation of citizens. In Kansas, youngsters use interactive map analysis tools to follow bird migration patterns. The states of Kansas, Illinois, and Massachusetts have particularly good websites (see Internet Resources).

Central to the success of student GIS explorations is the availability of map-based information and datasets, and these are becoming much more widely accessible as governmental offices, business, and academic institutions use GIS more. Public domain datasets are now widely available within each state and in many cities, and of course, much is available on the internet as well. ESRI works with a growing community of data providers in its ArcData Publishing Program, and educators can access a range of physical and human geography topics that include climate information, census data, and environmental data (see Figure 2). To make your early GIS explorations simple, you can even do all your GIS data exploration online with a standard internet browser without any special plug-ins or applets. ESRI sponsors the Geography Network (*www.geographynetwork.com/maps/arcexplorerweb.html*) where you can access ArcExplorer Web, a custom web mapping application.

Figure 2 ArcData online

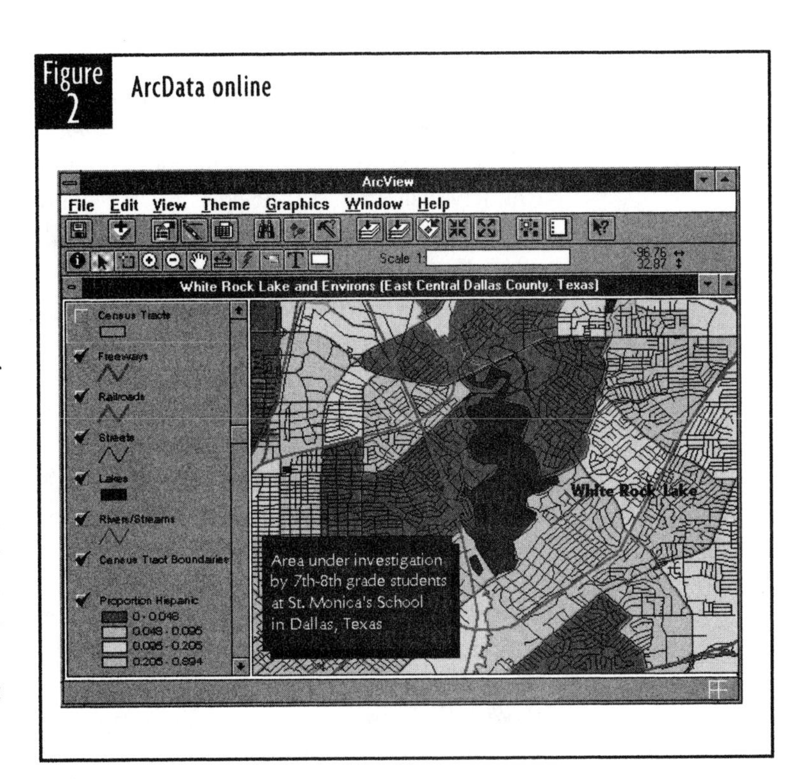

Applications

In middle school, students begin examination of special topics and regions. They often study a given phenomenon over space, seeing how it relates to others. They can survey the characteristics and relationships of geographically varied traits (population, economic makeup, and physical features). They can examine a region and its features and begin to understand the complex traits that identify and unite or separate areas, at scales from local to global. Middle school is also the best opportunity for engaging in cross-disciplinary studies.

Students in grades 5–8 have built up life experiences that can be used to their advantage. For example, their existing computer skills make it easy to involve them in producing geographic data about the local area. They can map trees in the neighborhood, parks and recreation centers, empty houses, safe bicycle routes, potholes in the roads, and so on. By creating GIS data that the middle school community can actually use, students learn the importance of accuracy in measurement, documentation, and representation (Baker and Case 2000).

Not all teachers will want to start with full-blown GIS software projects; however, there are many options involving sophisticated map projects. One of the richest listings of websites related to mapping is available from the University of Iowa's Center for Global and Regional Environmental Research (*www.cgrer.uiowa.edu/servers/servers_references.html*). Here you will find page after page of websites listed that relate to mapping in one way or another. For free topographical maps, one of the first places to visit is the U.S. Geological Survey website (*www.usgs.gov*). Also, many commercial enterprises and numerous other governmental agencies maintain maps and map-related data that your students can use. Especially powerful are a whole new array of interactive maps (many in Java) that draw maps showing the details that you request. To get an idea of the potential of these technologies, visit the GIS section of the website of the Environmental Protection Agency (*www.epa.gov/region02/gis/index.html*), which explores ways to integrate GIS and scientific visualization technologies to assist with environmental sciences decision making. The EPA site also offers GIS datasets for download at *www.epa.gov/airmarkets/cmap/data/index.html*.

One of the greatest free sources of maps and data is the U.S. Census Bureau, which can be accessed through several website sources. The bureau itself offers an interactive mapserver known as the Tiger Map Server, capable of depicting a multitude of layers, including highways, railroads, and congressional districts (*http://tiger.census.gov/cgi-bin/mapbrowse-tbl*).

As the 2000 census data are assembled, an important source of information will be the U.S. Census Bureau website American Fact-Finder (*http://factfinder.census.gov*), where you can find datasets containing information relative to any of the following: ancestry, citizenship, disability, educational attainment, income, industry, language spoken at home, marital status, migration, occupation, place of birth, place of work, poverty, rent, school enrollment, and more.

Another comfortable place to begin with students who are new to the world of maps is the MapTech mapserver (*http://mapserver.maptech.com*). At this site, you can request a topographical map of any major city in any state to view; you can also switch to an aeronautical chart or to an aerial photo or to a nautical chart (if appropriate) of that same area. As you move your curser across each displayed map, the latitude and longitude of that exact point is displayed. An equally inviting website, called MapMachine, is sponsored by the National Geographic Society (*http://plasma.nationalgeographic.com/mapmachine*). This interactive site also produces maps of any location in the world; additionally, you can select street maps, atlas maps, and historical maps. Physical science teachers will appreciate the theme choices for maps showing soils, minerals, forests, land use, as well as the ecoregions of the world. Moreover, under a link, you can ask your students to find out more about GIS and download GIS viewer software and example datasets relative to selected cities. But the mother of all mapservers is probably Microsoft's TerraServer (*http://terraserver.homeadvisor.msn.com*), which holds one of the world's largest online databases, providing free public access to a vast data warehouse of maps and aerial photographs of the United States. The views of cities are typically fascinating. TerraServer works with any computer system likely to be found in a school, and browsers even work over slow speed communications links. Your students can do advanced searches and even identify well-known places on the photos.

The use of PDA devices has become commonplace, and you might want to combine such handheld devices and related GPS technology as an added component in your instruction (see Figure 3). It is possible to download maps of local areas that are rich in geographical details, and these, in turn, can be downloaded to PDA devices for further navigational exploration. With the PDA navigational software Fugawi, it is possible to load U.S. Geological Survey topographic maps of any area, including the area surrounding your school, into the handheld's map library, and you can then ask your students to navigate to key geographical points on that map. Students can use the route function of the GPS and watch their progress over the terrain of the map on the screen. You can devise endless variations on treasure hunts or orienteering-like activities.

GIS and assessment

Not only can GIS facilitate performance-based assessments that stress the collection, organization, and integration of information; GIS illustrations can be used to measure complex academic achievement through the use of the interpretive exercise question. For example, with GIS illustrations, an interpretative exercise question can be designed as a series of multiple-choice questions. Students would need to provide responses at Bloom's analysis level, which requires them to divide a whole into its component elements. Therefore, middle school science teachers will have a greater range of flexibility in measuring the higher levels of learning, such as ability to recognize inferences, assumptions, and relevance of information.

Conclusion

Clearly, the incorporation of GIS-based science activities adds to the kaleidoscopic menu of technology-rich activities that are available for middle school students in these modern times. Many science teachers across the nation are contributing to the fabric of technology-based learning and instruction; and their efforts will spawn additional discussion concerning GIS applications for middle school science teachers to implement into 21st century classrooms.

National Standards

Geographic Information System (GIS) classroom applications are aligned with the National Science Education Standards, specifically to Content Standard D: Structure of the Earth System, i.e., "Land forms are the result of constructive and destructive forces;" and Content Standard E: Science and Technology, i.e., "...students' work with scientific investigation can be complemented by activities that are meant to meet a human need, solve a human problem, or develop a product..." for students in grades 5–8 (NRC 1996).

Internet resources for GIS

Bureau of Land Management:
ftp://ftp.blm.gov/pub/gis/gis_db.txt
ESRI's Schools and Libraries Program:
www.esri.com/industries/k-12
Association for Geographic Information:
www.geo.ed.ac.uk/agidict/welcome.html
Volusia County, Florida:
http://volusia.org/gis/whatsgis.htm
Guide to Geographic Information Systems:
www.gis.com
KanGIS K–12 GIS Community:
http://kangis.org
Illinois:
www.gisillinois.org/index.html
MassGIS:
www.mass.gov/mgis/gisedu.htm

Figure 3

Figure 3. PDA-based GIS, ArcPad 6.0

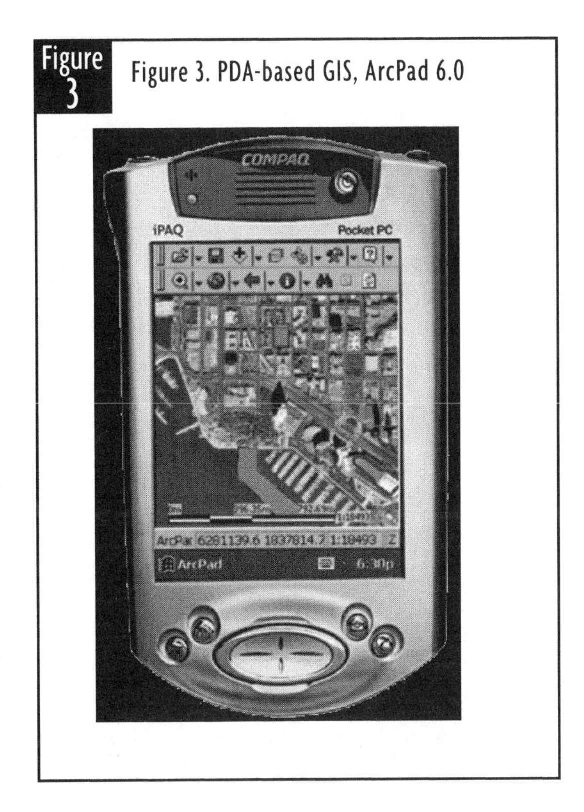

References

Audet, R. and G. Ludwig. 2000. *GIS in schools.* Redlands, CA: ESRI Press.

Baker, T. R. and S. B. Case. 2000. Let GIS be your guide. *The Science Teacher* 67(7) 24–26.

Davis, D. 2000. *GIS for everyone.* Redlands, CA: ESRI Press.

International Society for Technology in Education (ISTE). 1998. *National educational technology standards for students.* Eugene, OR: ISTE Press.

Underwater Web Work

MERVYN J. WIGHTING, ROBERT A. LUCKING, AND EDWIN P. CHRISTMANN

Looking for ways to enhance your oceanography unit? Consider the many online resources available to help you explore the mysteries of the deep. The following is a collection of sites appropriate for middle level classrooms.

An exciting beginning point is the website provided to educators by PBS, *www.pbs.org/oceanrealm/intheschool/index.html*. The website was developed as a companion to the PBS series Secrets of the Ocean Realm (see Resources). Series highlights included opalescent squid that rise from the abyss when night falls, strange creatures that inhabit cathedral-like forests of giant kelp, and never-before-seen behaviors of sharks, whales, dolphins and other more unusual marine dwellers like scorpion fish and wolf eels. The accompanying science activities for teachers available at the website were developed for use with students in grades 5–7 (with extensions for lower and higher grades). The activities explore such topics as oceanography, marine biology, ecology, physics, and conservation. Units of study are arranged around standard curricula and include objectives, background information, a list of materials, procedures, follow-up evaluation, and some additional internet resources.

The American Meteorological Society's website, *http://64.55.87.13/amsedu/DS-Ocean/news.html*, developed in conjunction with the National Oceanic and Atmospheric Administration (NOAA), is perfect for finding current events and news items that relate to oceanography topics that are themselves on the internet. The actual sources include CNN, BBC, science publications, and governmental agencies, so with a single click you can access or print a timely news item to share with your class that day.

NOAA has more in its treasure chest at *http://64.55.87.13/amsedu/DS-Ocean/home2.html#educ*, where it provides a wealth of information relative to oceanographic matters. Among the efforts they highlight is the National State Sea Grant, which is active in every coastal state and those surrounding the Great Lakes. The National Sea Grant College Program encourages the wise stewardship of our marine resources through research, education, outreach, and technology transfer, and Sea Grant is a partnership between the nation's universities and NOAA. The National Sea Grant Library has its own website, *http://nsgd.gso.uri.edu/diglib.html*, which is searchable and allows visitors to download documents as PDFs.

The Bridge website (*www.vims.edu/bridge*), features an Ocean Sciences Teacher Resource Center, which contains an array of classroom materials. For example, you can find a collection of lesson plans designed specifically for grades 6–8 on topics such as El Niño, ocean exploration, marine mammals, and salinity. The Data Activities section contains an inventory of data-related lesson topics that can be used in the classroom. Examples include Diversity of the Deep, the Far-Reaching Effects of Oil Spills, and the Scoop on Scallops. Activities are organized by subject: biology, human activities, ecology, physics, chemistry, climate, and geology. The site also gives access to real-time data from a variety of receiving, recording, and transmitting instruments through GoMOOS, a national pilot program that provides hourly data transmissions from sources in the Gulf of Maine. The data stream includes wave height and current strength, and students can use it to create their own database, plot data, and make their own forecast predictions.

Another resource that will bring the underwater world into your classroom free of charge can be found by registering with Project Oceanography at *www.marine.usf.edu/pjocean*. This site provides information about a live television program designed for middle school science students. Each week during the school year, students can learn about a variety of ocean science topics that are taught by real scientists, and the program features a call-in question-and-answer session at the end of each lesson. Teacher packets are supplied online, and they include information relating to program topics, activities for students, and student information sheets. Topics have included marine debris, the sounds of the seas, and the world of plankton.

An additional site that will capture the minds of middle school students is Ocean Explorer, found at *http://oceanexplorer.noaa.gov/explorations/explorations.html*. Available at this site are lesson plans for students that are specifically tied to deep-water exploration projects. These lesson plans focus on cutting-edge ocean exploration and research, using state-of-the-art technology aboard *Discovery*, one of the nation's most sophisticated research vessels. The site incorporates lesson plans developed for grades 5–6 and 7–8. Lessons include biodiversity of deep-sea corals, positive and negative buoyancy in the ocean, and the impact of dumping waste.

Every science teacher looking for oceanography resources will love to visit the Birch Aquarium—and so will students! The site is the interpretive center for the Scripps Institution of Oceanography in La Jolla, California, and it can be accessed at *www.aquarium.ucsd.edu*. Click the Learning Center button to explore topics on sharks, cuttlefish, or seahorses, and explore the Featured Links to view archived footage of discoveries made while diving in the Pacific Ocean. There are also interactive connections that your students will enjoy, including the seahorse anatomy game, the seahorse environment, and "kelp kitchen."

Another resource at this site is Science Features. Here you will find a host of links that can be used with middle school students. We particularly liked FLIP, which stands for Floating Instrument Platform. It is actually a huge specialized buoy that gathers data on topics such as the way water circulates, how storm waves are formed, and the sounds made by marine animals. The site includes a hands-on project called "FLIP in a Bucket" that will enable students to replicate the design of the platform in your classroom (*http://aquarium.ucsd.edu/learning/learning_res/voyager/flip/flip_exp_print.html*). You can also visit another Scripps resource, Wyland Ocean Challenge (*www.wylandoceanchallenge.org*), to access interdisciplinary art and science educational material, including free downloadable teacher activities.

Finally, if any of your students are interested in exploring a career in oceanography, send them to *www.marinecareers.net/fields.htm*. This site provides information on careers in marine biology, oceanography, marine engineering, and other related fields, together with profiles of people who are practicing these careers. Some of these people working in the field supply an e-mail address and actively encourage students to write to them to ask questions about the profession. Another good career site that focuses on the achievements of women in oceanography is *www.womenoceanographers.org*. The site profiles a number of women engaged in a variety of oceanographic careers (e.g., Lauren Mullineaux, a scientist who studies animals that live on the seafloor, and Kathryn Gillis, a professor who dives to rifts in the seafloor (some as deep as six kilometers!) to learn about the processes taking place within the ocean crust. All of your students—girls and boys—will enjoy finding out more about the lives of these underwater explorers. Dive in!

National Standards related to oceanography

Content Standard D: Grades 5–8

- Water, which covers the majority of the Earth's surface, circulates through the crust, oceans, and atmosphere in what is known as the "water cycle." Water evaporates from the Earth's surface, rises and cools as it moves to higher elevations, condenses as rain or snow, and falls to the surface where it collects in lakes, oceans, soil, and in rocks underground.
- Water is a solvent. As it passes through the water cycle it dissolves minerals and gases and carries them to the oceans.
- Global patterns of atmospheric movement influence local weather. Oceans have a major effect on climate, because water in the oceans holds a large amount of heat.
- The Sun is the major source of energy for phenomena on the Earth's surface, such as growth of plants, winds, ocean currents, and the water cycle. Seasons result from variations in the amount of the Sun's energy hitting the surface, due to the tilt of the Earth's rotation on its axis and the length of the day.

Resources

Secrets of the Ocean Realm, PBS Home Video, available through *www.pbs.org.*

Reference

National Research Council (NRC). 1996. *National science education standards.* Washington, DC: National Academy Press.

Up-to-the-Minute Meteorology

MERVYN J. WIGHTING, ROBERT A. LUCKING, AND EDWIN P. CHRISTMANN

If you are looking for up-to-the-minute weather information, chances are you'll turn to The Weather Channel or visit a website such as *www.accuweather.com*. You can also turn to these resources to help teach your students meteorology. Of course, students should still be taught how to gather their own data using hands-on instruments such as thermometers, barometers, weather vanes, wind gauges, and rain gauges. Once they have mastered traditional data collection techniques, however, introduce them to the wealth of data, lesson plans, and simulations offered online and on television. Here are a few of our favorite weather-related resources.

Weather on the web

The Weather Classroom (*www.weather.com/education*), available from The Weather Channel, is a comprehensive collection of weather-related internet links for teachers. Once you set up a free teacher account, you can access lesson plans, interactive multimedia, newsletters, and caches of other teaching resources.

For example, a thematic lesson titled RAYS Awareness gives you a video clip from Weather Channel meteorologist Jennifer Lopez and four Standards-based lesson plans (Hot Colors, Light Reading, A (Time) Piece of Pizza, and Seed vs. Seed) that can be used with your middle school students. The unit and activities are designed to focus on the dangers of ultraviolet radiation to skin. The lessons include scientific experiments, creative writing exercises, mathematical calculations involving Sun protection factors (SPF), and explorations of how various cultures and economies are affected by the Sun.

In addition to the variety of weather units and lessons available for classroom use, teachers can tape episodes of The Weather Classroom, which airs Monday and Thursday from 4–4:30 A.M. EST. Topics of episodes have included winter weather, careers in meteorology, weather and geography, and hurricanes. A teachers guide for each episode created by veteran teachers can be found on the website. Each guide contains background information, activities, and study questions. Additionally, some of these guides contain sophisticated data-gathering forms and matrices.

Another offering on The Weather Channel site is Project SafeSide (*www.weather.com/safeside*), developed in partnership with the American Red Cross. SafeSide explores the effects of extreme weather such as hurricanes, tornadoes, floods, lightning, extreme heat, and winter storms. One part is designed for grades K–4 and the other for grades 5–12. Both include hands-on activities, cross-curricular extensions, and instructions for conducting a severe weather simulation. Students are encouraged to develop a family disaster plan to protect and prepare themselves and their families for a weather emergency.

Finally, encourage your students to visit *www.weather.com/learn/dave*, where they will find a

Figure 1 Climograph image via MetLinkInternational

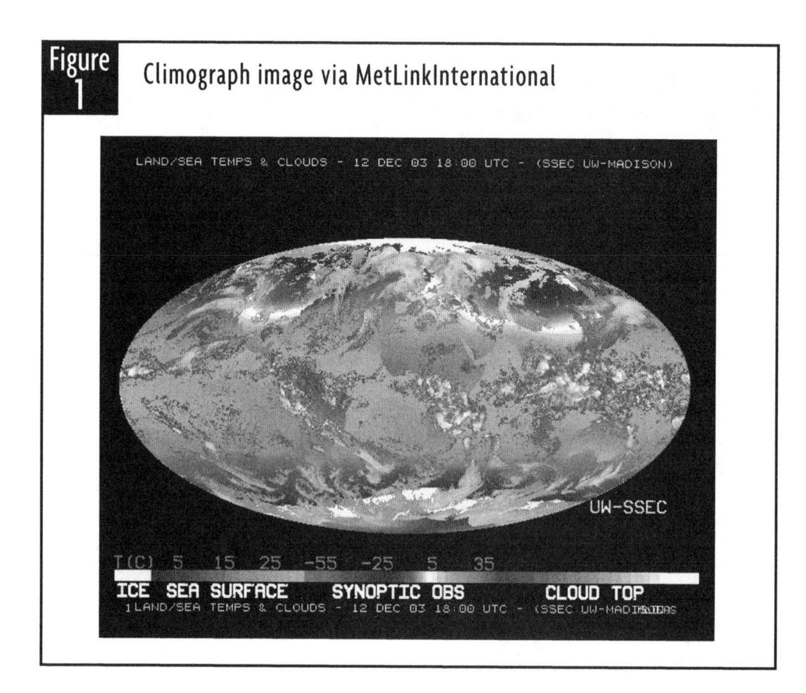

wide range of weather-related information for incorporation into their assignments.

Global perspective

If you are interested in global meteorology, check out the MetLinkInternational Weather Project at *www.metlink.org*. This project is the result of a partnership of The Royal Meteorological Society, Project Atmosphere Australia, and MetLinkInternational. Students around the world use the site to exchange weather observations using an online database. For example, students in the United States can exchange weather information with classes in Oman, India, and Australia. Teachers can find additional project suggestions from participating schools around the world.

Daily images of the entire globe are provided that show cloud movement and other atmospheric events using still photos, movies, and weathercams. With the click of a mouse, students can see fog surrounding the Eiffel Tower or a sunrise in Hong Kong. Figure 1 is an example of a Climograph image, which shows worldwide temperatures.

If these images capture your imagination, try the weather and climate site provided by NASA at *www.jpl.nasa.gov/earth/weather_climate/weather_climate_index.cfm*. Here you will find animated images of Hurricane Isabel that were generated using data collected by the Atmospheric Infrared Sound experiment aboard NASA's *Aqua* spacecraft. The "Sounder" collects highly accurate measurements of air temperature, humidity, clouds, and surface temperature, which are then compiled to create false-color imagery of oceanic and atmospheric events. You can also watch a clip of the changing isotherms of a developing storm as it crosses the Atlantic from the west coast of Africa. For another fascinating set of images, check out the many photos of lightning at *http://thunder.msfc.nasa.gov* (see Figure 2). At this link, your students can observe data on the global frequency and distribution of lightning as observed from space by NASA's Optical Transient Detector.

Grab bag

Another useful site is *www.education.noaa.gov*, home of the National Oceanic and Atmospheric Administration's education resources. Here you will find lists of free and inexpensive classroom materials, such as booklets, videos, software, and posters. You'll also find a collection of K–12 weather and climate activities and study guides on weather topics. One in particular that caught our attention was the Global Weather Services vision of weather in the distant future, which can be viewed at *www.ucar.edu/pres/2025/web*. At this site, researchers project weather patterns for the year 2025 using video images of real and model data.

Flight into the fury

From the safety of your classroom, students can join the crew of the 53rd Weather Reconnaissance Squadron and take an interactive flight into the eye of a hurricane by visiting *www.*

hurricanehunters.com. This site has excellent videos, an interesting narrative, and plenty of pictures of storm systems. Another site that your students can use for investigating severe weather is *www.tornadoproject.com*, which gives students scientific information about tornadoes and explains some of the hazards associated with them.

Conclusion

Teachers need to take full advantage of the ubiquitous, up-to-the-minute information that is available at our fingertips. Otherwise, students might not make the connections between school and our increasingly technological world.

Internet Resources

AccuWeather:
    www.accuweather.com
Weather Classroom:
    www.weather.com/education
Project SafeSide:
    www.weather.com/safeside
Dave's Weather Dictionary:
    www.weather.com/learn/dave
MetLinkInternational Weather Project:
    www.metlink.org
NASA's Weather and Climate site:
    www.jpl.nasa.gov/earth/weather_climate/weather_climate_index.cfm

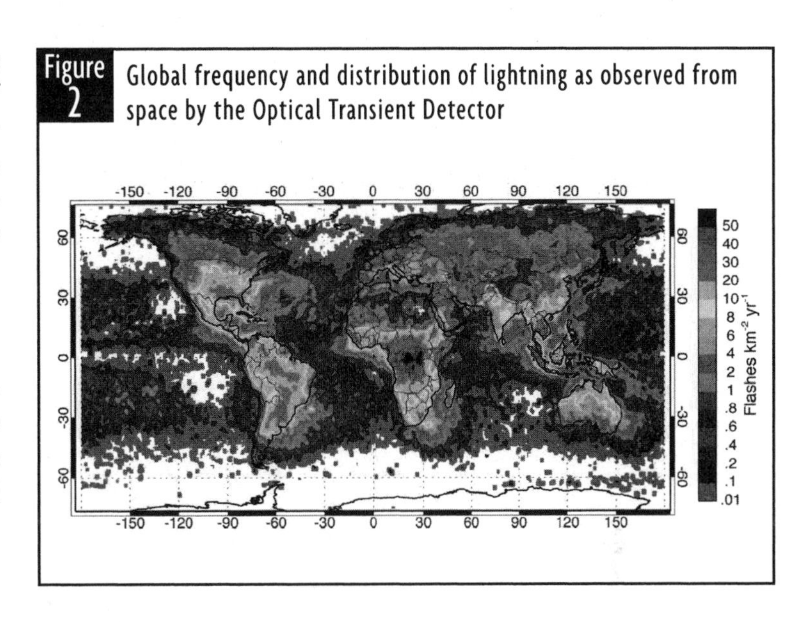

Figure 2 Global frequency and distribution of lightning as observed from space by the Optical Transient Detector

Lightning and Atmospheric Electricity:
    http://thunder.msfc.nasa.gov
NOAA's Education Resources:
    www.education.noaa.gov
Global Weather Services:
    www.ucar.edu/pres/2025/web
The Hurricane Hunters:
    www.hurricanehunters.com
Tornado Project Online:
    www.tornadoproject.com

Technology-Based Planetary Exploration

JOHN HARRELL, EDWIN P. CHRISTMANN, AND JEFFREY LEHMAN

Too often in science classes, the textbook becomes the sole deliverer of scientific knowledge to the passive student. Consequently, the student develops the notion that science is merely a vast compendium of scientific facts, instead of a dynamic process of discovery and learning. The National Science Education Standards prescribe a change in the emphasis of science education from factual memorization and textbook learning to a more exploratory approach (NRC 1996). A shift toward inquiry-based learning is one of the main tenets of the new approach to science education, which also includes technology and the history/nature of science as two other tenets.

As more and more schools acquire access to the internet, science teachers have a responsibility to students to use it to elaborate upon scientific concepts that are often treated superficially in the textbook. The amount of data that can be acquired from the internet is staggering, and students need experience gathering, manipulating, and applying these data. By allowing students to actively seek out information and determine what facts are relevant, the interactive use of the internet can help students become active learners.

The use of the internet in science classes provides a database for students for many tasks. To use the internet successfully, students must acquire data-retrieval skills. As material is retrieved from the internet, students must handle it both effectively and efficiently. Specifically, students must select data that are pertinent to the task at hand by thinking critically about the relative importance of material. The lesson that follows is an example of how science educators can incorporate the internet into the classroom.

Standards alignment

The National Science Education Standards affirm that middle school science subject matter should focus on science facts, concepts, principles, theories, and models. Also, one of the goals of the Standards is for science teachers to synthesize the history and nature of science into an understanding of the natural world. Moreover, authors of the Standards have outlined the Earth and space science unifying concepts for grades 5–8 (NRC 1996). Conformably, this planetary exploration, which is designed for middle school students, focuses on Middle School Content Standard D, the Earth's history, and also emphasizes Earth's position in the Solar System.

Additionally, this technology-based planetary investigation incorporates a number of performance indicators outlined in the National

Figure 1 A plot of the average distance of each planet from the Sun

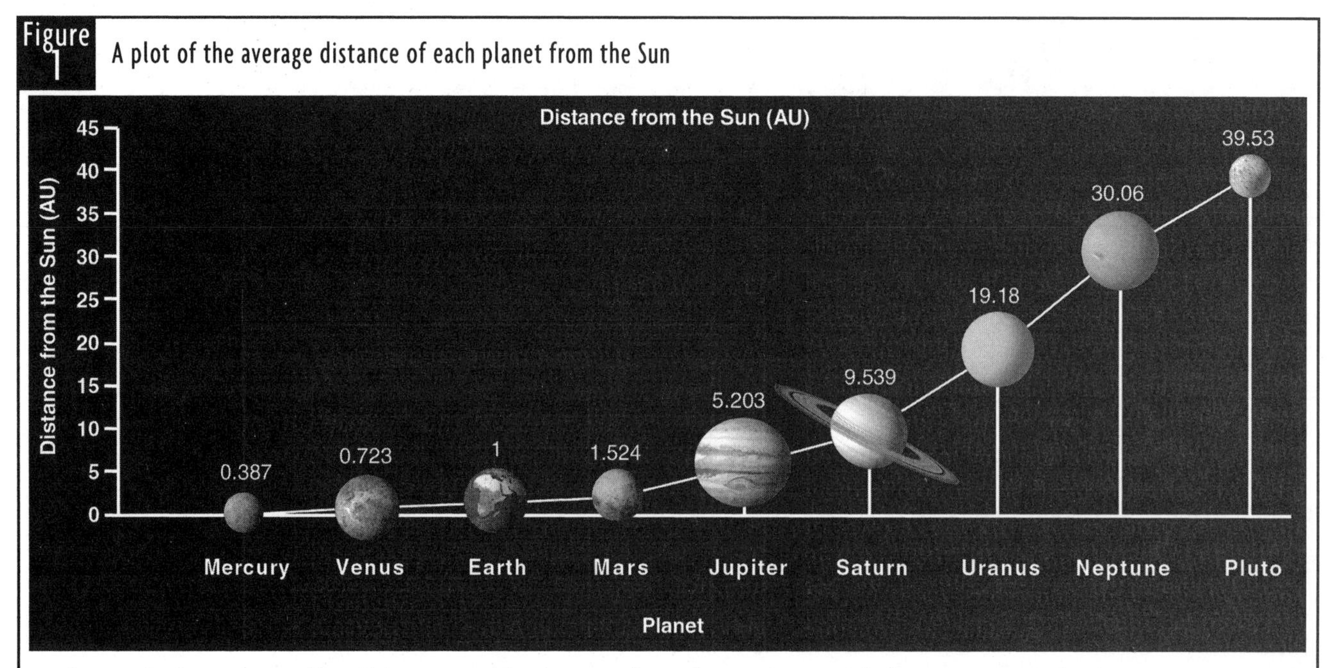

In the graph above, a trendline for one variable, distance from the Sun, is provided. By examining the representation, students can be asked to consider why scientists often group these planets into two categories, those close to the Sun and those farther away. Once the students realize how the planets are grouped, questions concerning the attributes of the planets that are grouped together should arise. This provides an excellent lead-in to a discussion of the similarities and differences that exist between the Terrestrial planets (near the Sun) and the Jovian planets (far from the Sun.) Why are all the denser planets closer to the Sun? Why are planets that are farther from the Sun composed of light gases? Can we divide the planets into distinct categories? Similarly, each crew can be assigned to examine patterns or trends from class data for a different variable from the class database; e.g., period of rotation.

Technology Standards for middle school students. For example, students will select and use appropriate tools and technology resources to accomplish a variety of tasks and solve problems. Also, students will collaborate with peers, experts, and others using telecommunication and collaborative tools to investigate curriculum-related problems, issues, and information, and to develop solutions or products for audiences inside and outside the classroom (ISTE 1998).

In order to accomplish this, students break into crews to explore one of the nine planets in the Solar System. Each group is responsible for gathering data, such as the planet's average distance from the Sun, atmospheric composition, average density, radius, mass, surface composition, period of rotation, and the period of revolution of the planet assigned. These data can be found at various sites on the internet. See Internet Resources for a list of sites that are colorful, easy to navigate, and provide hot links to similar sites. A plethora of planetary pictures and diagrams pertaining to planetary motion can also be accessed at these sites.

SCI LINKS. THE WORLD'S A CLICK AWAY

Explore scale, powers of 10 at *www.scilinks.org.* Enter code SS30101.

Another option is to have the students conduct their own searches on the internet. Students

Figure 2 String lengths for solar system model

Planet	Mercury	Venus	Earth	Mars	Jupiter	Saturn	Uranus	Neptune	Pluto
String length (m)	0.39	0.72	1.00	1.52	5.20	9.54	19.18	30.06	39.53

who conduct individualized searches will have to think critically about the source of the information. This alerts students to the fact that not all scientific data are completely accurate, and the reliability of the source must be taken into consideration when selecting data. Once the necessary data are obtained, they should be compiled in a class database and analyzed by the students. One option here is to have students determine whether data obtained can be verified by multiple, independent sites—a worthy scientific task in itself.

To help with the analysis, students can use spreadsheet software with graphing capabilities similar to Microsoft Excel or graphing calculators to manipulate the data obtained from the search. While plotting the data by hand is possible, it is often a tedious and time-consuming task that is not the objective of this lesson.

Data plotted with user-friendly spreadsheet software and graphing calculators have many advantages over data presented in a tabular format. One such advantage is the ability to create trendlines using the data points. Trendlines illustrate the relationship between each data point on the graph, and can be given a mathematical representation in the form of an equation of the line (see Figure 1). This demonstrates the critical link between mathematics and science.

Another activity that would be beneficial to the students would be to create a model of the solar system, thus integrating data obtained using technology with a concrete visual experience. However, models can produce misconceptions if they are not done to scale (Debuvtis 1990). Furthermore, when students are taught with models, they often only learn the model and ignore the concept (Dyche 1993). To avoid instilling science misconceptions, such as the common notion that the planets are aligned, care must be taken when constructing the model.

One model activity uses students, a football field, and a lot of string. The students may be left in their previously formed crews for this activity. Each crew is responsible for measuring and cutting the length of string that is representative of the orbital distance of their planet in astronomical units, which are units equal to the orbital distance of Earth (see Figure 2). To heighten the effect, encourage students to wear clothing that is the same color as their planet. For example, the Mars crew would wear red and the Earth crew would wear blue and white.

References

DuBuvitz, W. 1990. The importance of scale drawings or: Let's not blow things out of proportion! *Physics Teacher* 28(9): 604–605.

Dyche, S, P. McClurg, J. Stepans, and M. L. Veath. 1993. Questions and conjectures concerning models, misconceptions, and spatial ability. *School Science and Mathematics,* 93(4):191–197.

International Society for Technology in Education (ISTE). 1998. *National educational technology standards for students.* Eugene, OR: Author.

National Research Council (NRC). 1996. *National science education standards.* Washington, DC: National Academy Press.

Internet Resources

The nine planets: A multimedia tour of the solar system: *http://seds.lpl.arizona.edu/nineplanets/nineplanets/nineplanets*

Solar system exploration: A multimedia tour of the solar system: *http://solarsystem.nasa.gov/planets/index.cfm*

How Reliable Is the Temperature Forecast?

EDWIN P. CHRISTMANN

Project 2061 suggests "technology provides the eyes and ears of science—and some of the muscle too. The electronic computer, for example, has led to substantial progress in the study of weather systems..." (1990). Obviously, now that teachers have access to a kaleidoscope of technological advancements, middle school science teachers can engage students in classroom activities that are inquiry-based and technology-rich.

The internet has become as commonplace as a daily newspaper and is a primary source of news for middle school students. Access to up-to-the-minute weather forecasts is something most students now take for granted. However, like most of us, middle school students probably don't question the accuracy of what they read online. This activity is designed to let students assess the accuracy of internet-based weather forecasts through technology-based inquiry.

This activity can be done with five computer stations, with students assigned into laboratory groups of four students. However, if only one computer is available, the activity can be done as a demonstration for an individual class. Moreover, teachers who have several classes throughout a school day can alternate days for classes so that period 1 conducts their laboratory on Monday, period 2 conducts their laboratory on Tuesday, and so on. After the data collection is complete, teachers can have students analyze the data collected by other classes.

Note: I refer to a particular brand and model of probe in this article so I can provide detailed directions and discuss capabilities in detail. However, you can use any temperature probe for this activity and modify the procedures to fit the capabilities of the hardware and software available to you. (See Resources for a list of alternate temperature probes.)

Temperature Reality Check:

Problem statement

To introduce inquiry as part of the scientific process, we will ask the question, "Is the temperature forecast provided by *www.weather.com* reliable?" Students can answer this question by conducting a scientific investigation and reporting the findings.

Materials

- Computer (preferably a laptop) with internet access
- Go! Temp Probe (available for $39 through Vernier)

Table 1 weather.com predicted readings

Date	Time	Temp
10/18/04	9:00 p.m.	8°C
10/18/04	10:00 p.m.	8°C
10/18/04	11:00 p.m.	8°C
10/19/04	12:00 a.m.	8°C
10/19/04	1:00 a.m.	8°C
10/19/04	2:00 a.m.	9°C
10/19/04	3:00 a.m.	9°C
10/19/04	4:00 a.m.	9°C
10/19/04	5:00 a.m.	9°C
10/19/04	6:00 a.m.	9°C
10/19/04	7:00 a.m.	9°C
10/19/04	8:00 a.m.	9°C
10/19/04	9:00 a.m.	9°C
10/19/04	10:00 a.m.	9°C
10/19/04	11:00 a.m.	11°C
10/19/04	12:00 p.m.	11°C
10/19/04	1:00 p.m.	12°C
10/19/04	2:00 p.m.	12°C
10/19/04	3:00 p.m.	12°C
10/19/04	4:00 p.m.	12°C
10/19/04	5:00 p.m.	12°C
10/19/04	6:00 p.m.	12°C
10/19/04	7:00 p.m.	11°C
10/19/04	8:00 p.m.	12°C

- Logger Lite Software (free with Go! Temp Probe from Vernier)

Procedure

Go to *www.weather.com* and enter the zip code for your area. Next, select the day that you will be taking the readings at your school, for example, Tuesday. Also, make sure that you have clicked "Show this page in: metric units." This will be used for comparison to your recorded hourly temperatures. Beside "Hourly Forecast," click on "More Details," which will give you hourly temperature forecasts. You will have to click "Next" so that you have 24 hours of hourly temperature predictions.

Record the hourly temperatures that have been predicted by *www.weather.com*. For example, Table 1 shows the temperatures that *weather.com* predicted for Slippery Rock, Pennsylvania, from 9:00 p.m. on 10/18/04 through 8:00 p.m. on 10/19/04.

Load and activate your Logger Lite Software and install your Go!Temp probe into your computer. To program your Logger Lite Software: Under "Experiment" go down to "Data Collection" and program the following:

Mode: Time Based
Length: 24 Hours
60 samples/hour

Once your computer is programmed and ready to go, place it in a location that is secure, but still allows the attached probe to be placed outside. The Go! Temp probe cable is approximately seven feet long and about ½ cm thick, which should allow the probe to reach a window. Be careful, however, not to pinch the probe's cable while closing the window after you place the probe outside. Setting up the laptop and probe in another secure location in the building or asking a student to be in charge of the data collection in their home are also options. If the classroom is the only option but security is a problem, modify the activity to complete the data gathering for a single day or class period.

After you have programmed the software and situated your Go! Temp probe, go to your Logger Lite Software and under "Experiment," press "Start Collection." Your Logger Lite Software will take readings for the next 24 hours. Make sure that you begin taking readings at the exact same time that you started your predicted readings from *www.weather.com*. For example, I started my data collection through Logger Lite at 9:00 p.m. on 10/18/04.

As you are taking readings, you can see the current temperature on the desktop by viewing the spreadsheet, graph, and thermometer on-

Figure 2 Viewing the data as a spreadsheet, thermometer, and graph

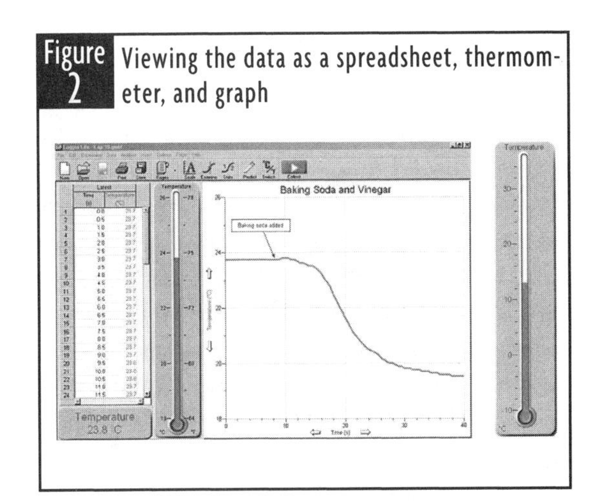

Figure 4 Logger Lite's statistical feature

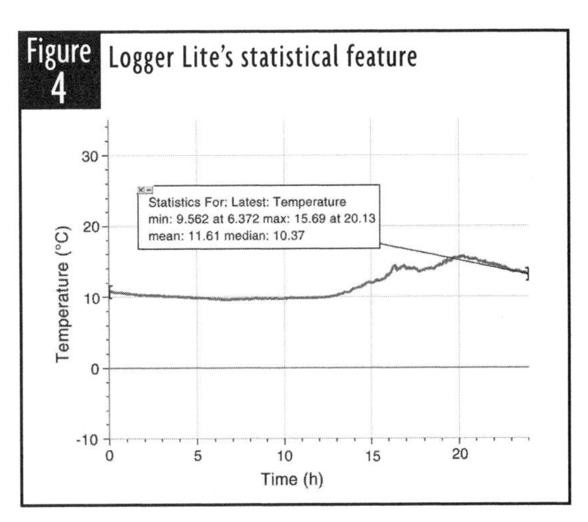

Figure 3 Temperature readings recorded with Logger Lite for 24 hour Time (h) 0 corresponds with 9 p.m. on 10/18/04. Subsequently, Time (h) 24 corresponds with 8 p.m. on 10/19/04.

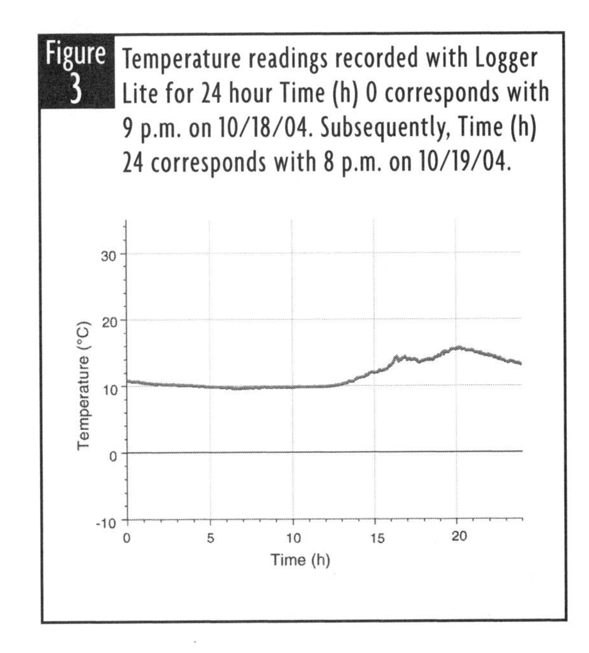

screen (Figure 2). You can share the screen with the entire class by using a projector (see page 101 for more information on projectors). Another alternative is to print out screenshots for students as the activity progresses.

Once you have finished data collection, under "Experiment," select "Store Latest Run." This will store all of your data collection. Figure 3 shows my data-collection readings over 24 hours.

Calculate statistics and speculate

Logger Lite has a built-in statistics calculator. Under "Analyze," if you select "Statistics," students will have the mean and median temperature, as well as the minimum and maximum temperatures (Figure 4). Therefore, students can analyze the range of temperatures over the 24-hour span. Student analysis of the data can lead to questions they can investigate, including:

- When was there less variation in the reading? What reason(s) can you suggest for steadier readings at this time?
- When did the low and high temperatures occur? What factors contributed to these readings?
- From differences you see in the temperature readings during the day, what can you infer about the amount of sunlight on the thermometer? Provide observations and reasoning to support your inferences.
- Did cloud cover affect temperature during the day? If so, how?

If the proper instruments are available, students can also explore how temperature is affected by atmospheric pressure, precipitation, wind, and other meteorological factors.

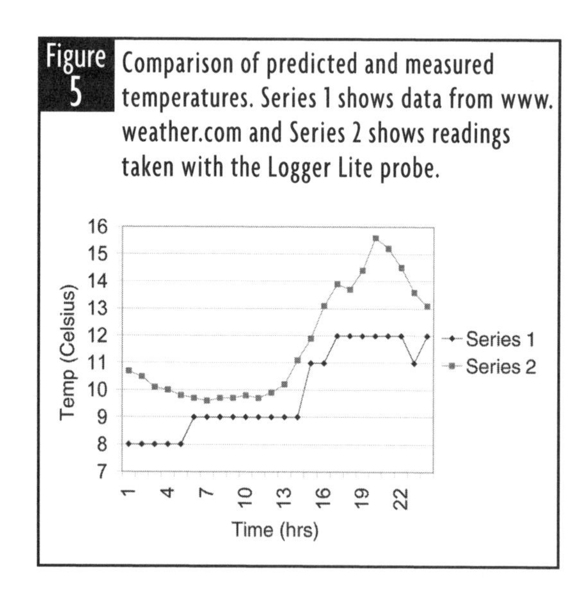

Figure 5 Comparison of predicted and measured temperatures. Series 1 shows data from www.weather.com and Series 2 shows readings taken with the Logger Lite probe.

Analysis

To answer our original question, "Is the temperature forecast provided by *www.weather.com* reliable?," students can plot the data from Figures 1 and 3 to compare the predicted temperatures for the 24-hour period to the actual temperatures. In my data, the temperatures forecasted by *www.weather.com* were slightly lower than the temperatures that were measured using the probes; however, the pattern is very similar. In essence, reliability is the consistency of the results. Students establish reliability by comparing the consistency of the patterns between the predictions by *www.weather.com* and the measurements taken with the probes.

In this experiment, although the forecasted temperatures are slightly lower than the actual measurements, the graphical pattern of increases and decreases throughout the day is very consistent. For the statistically savvy teacher, another way to check the consistency of the results is to compute a Pearson correlation coefficient. In this case the Pearson correlation coefficient showed $r = .918$, $p < .01$, which means that we are 99% confident that there is a very strong relationship between the two sets of data.

Conclusion

Having students collect temperature data through technology-based inquiry is commensurate with tenets of the National Science Education Standards (NRC 1996). This activity, which is based on Content Standards A and B, demonstrates that temperature data gathered with the proper tools are more accurate than the data predicted by weather forecast. It also teaches students how to use the latest technology to gather data in an inquiry setting. In addition, the interactive graphs that can be generated by Logger Lite software help students interpret the results of their experiments and can be used to create professional laboratory reports.

References

National Research Council (NRC). 1996. *National science education standards.* Washington, DC: National Academy Press.

Project 2061. 1990. *Science for all Americans.* Oxford, England: Oxford University Press.

Internet resources

Vernier Logger Lite:
http://elementary.vernier.com/products/logger-lite.html

Vernier Go! Temp:
www.vernier.com/go/gotemp.html

Alternate temperature probes

Labpro:
www.vernier.com/mbl/labpro.html

HOBO Data Logger:
www.onsetcomp.com/Whats_New/New_Products.html

CBL 2:
http://education.ti.com/us/product/tech/datacollection/where2buy/where2buy.html

ImagiProbe:
www.imagiworks.com/Pages/Products/ImagiProbe.html

Viewing Volcanoes

MERVYN J. WIGHTING, ROBERT A. LUCKING, AND EDWIN P. CHRISTMANN

When Mount St. Helens threatened to erupt again in 2004, it grabbed headlines and captured the imagination of the country. Science classrooms nationwide used the event as an opportunity to make real-world connections to Earth science concepts introduced in the classroom. Thanks to modern technology, you no longer have to wait for the next eruption to take advantage of your students' natural interest in volcanoes. Using a variety of media, you can now explore these forces of nature in a variety of ways—all from a safe distance!

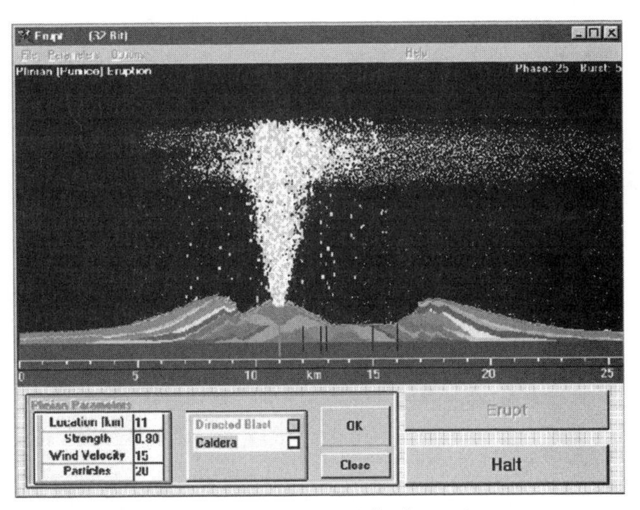

Simulate an eruption at www.ecs1.lanl.gov/wohletz/Erupt.htm

Volcanoes online

In AD 79, Mount Vesuvius near Naples, Italy, erupted and buried the towns of Herculaneum and Pompeii. The lava flowed so quickly and devastatingly that everything in its path was preserved. Excavations of these sites provide students with a glimpse of how Europeans lived almost 2,000 years ago. Visit *http://volcano.und.nodak.edu/vwdocs/volc_images/img_vesuvius.html* to be transported to Vesuvius. Follow up with a visit to *http://www.archaeology.org/interactive/pompeii/history.html* for an interactive archaeological dig among the ruins of Pompeii. The area surrounding Pompeii was hit by a tsunami, caused by the force of the eruption. Have students compare the cause and effects of this ancient tsunami with the one that hit in December of 2004.

To impress upon your students that volcanic eruptions are a part of everyday life, have them visit *www.fs.fed.us/gpnf/volcanocams/msh* to view a photo of the Mount St. Helens volcano taken within the last five minutes by the VolcanoCam permanently focused on the crater. The VolcanoCam is located at the Johnston Ridge Observatory, which is situated five miles from the volcano at an elevation of 4,500 feet. For other up-to-date information, visit *http://hvo.wr.usgs.gov/maunaloa/current/main.html* for a daily log of volcanic activity within the United States. Also, a dramatic slide show of Mount St. Helens is available at *www.msnbc.msn.com/id/6092368.* At this site you'll also find a link to an excellent interactive presentation called "The Anatomy of a Volcano" that explains why a volcano erupts.

Why would anyone want to live near a volcano? Students can find answers to this question by doing a lesson like the one found at *www.volcanolive.com/lesson2.html*, which examines some of the risks and benefits of living near volcanoes. For further study, look at *www.geo.mtu.edu/volcanoes*, a website that explains links between volcanoes and people. This website offers viewers a great deal of interesting data about the hazards of volcanoes, such as the effects of volcanic dust on aviation, as well as dozens of other volcano-related links.

To show your students some of the more well-known volcanoes, visit *www.cotf.edu/ete/modules/volcanoes/vtypesvolcan1.html.* This site offers a wealth of graphics and text that explain how volcanoes are formed and how and why they erupt. Your students will be captivated by the displays of types of lava, sizes of eruptions, and the volcanic eruption animations. For more information, visit Volcano World, a site that bills itself as being the web's premier source of volcano information, at *http://volcano.und.nodak.edu/vw.html.*

To develop an activity with a volcano simulation, you can download a simulation of volcano eruptions for free at *www.ees1.lanl.gov/Wohletz/Erupt.htm.* Undoubtedly, this will give you something more scientific than the traditional baking soda and vinegar simulation typically used with middle school students.

There are several excellent lesson plans on volcanoes designed for students in grades 4–8 that have been compiled by the United States Geological Survey. These lesson plans focus on the powerful eruption of Mount St. Helens in 1980. For example, Lesson 2 covers the actual eruption and has practical activities that simulate the events that occurred. Lesson 5 explains the long-term effects of a volcanic eruption and has activities that specifically explore how plant life is affected. The entire set of lesson plans can be found at *http://interactive2.usgs.gov/learningweb/teachers/volcanoes.htm.*

The Discovery School also makes available a set of volcano lesson plans for middle level students. These review the main types of volcanoes and provide instructions for modeling the different types in the classroom. These well-designed plans are available at *http://school.discovery.com/lessonplans/programs/understanding-volcanoes.*

Volcanoes in print

Books are another great way to dig into volcanoes. *Krakatoa: The Day the World Exploded—August 27, 1883*, by Simon Winchester, is a great teacher resource. It provides a vivid account of a volcanic eruption in the Indonesian arc, between the large islands of Sumatra and Java. The eruption of the volcanic island of Krakatoa triggered an immense tsunami that killed nearly 40,000 people. *Krakatoa* is also available on cassette or CD, and can even be downloaded for playing on personal audio devices. The books, CD, and audiotape are available at most bookstores. The audio file may be purchased and downloaded from *www.audible.com.*

Volcanoes onscreen

Several videos are available to help you bring volcanoes into the classroom. We recommend you explore the National Geographic's *Volcano!* For movies of Mount Etna on Sicily, visit *http://volcano.und.nodak.edu/vwdocs/movies/etna_mov.html.* Another option is the series of video clips made available by the commercial company Volcano Video Productions. Their website at *www.volcanovideo.com* allows students to download and play sample video clips of various scenes of eruptions of lava, although they are somewhat small in screen size.

The final medium we recommend is the CD-ROM *Volcanoes!* This teaching aid is a powerful multimedia reference that students will enjoy using to learn about volcanoes. It is packed with hundreds of full-color photos, computer animations, movie footage, sounds, factoids, and games that make learning about volcanoes fun. Students can use the CD-ROM in a self-directed activity to perform simulated experiments in a virtual laboratory, take video voyages (in either English or Spanish), or use its encyclopedia for further research. Alternatively, teachers can provide more direct instruction using the teacher's guide that comes with the CD-ROM; the whole package is available for $59.95 from *http://school.discovery.com.*

Conclusion

Having students explore volcanoes is supported by the National Science Education Standards. For example, students will demonstrate an understanding of scientific inquiry by simulating volcanic eruptions and analyzing data with computer software. The Physical Science Standard, Transfer of Energy, can be met by having middle school students recognize where magma reaches the surface of the Earth and by exploring volcanoes and the locations of fissures in the Earth. In addition, by having your students explain plate motions as a result of events such as volcanic eruptions, students will be better equipped to meet the tenets of Content Standard D: Earth and Space Science, which emphasizes an understanding of the Earth's History and the Structure of the Earth as a System. We hope the resources described in this article allow teachers to launch in-depth studies of volcanoes—far from any hot ash or lava flows.

Reference

National Research Council (NRC). 1996. *National science education standards.* Washington, DC: National Academy Press.

Resources

Volcanoes! Educational software ($20): *www.rockware.com*

From the Discovery Channel Teacher Store at *http://school.discovery.com*:

Discovery Channel video *Understanding Volcanoes* ($59.95)

Discovery Channel *Volcanoes Teachers A–Z Resource Guide* ($0.99)

Discovery Channel Volcanoes CD-ROM, v. 2.0 ($59.95)

Discovery Channel Volcanoes Video Quiz ($49.95)

Earth Science CD-ROM Collection ($549.95)

Volcanoes: Mountains of Fire video ($49.95)

Websites

Mt. Vesuvius: *http://volcano.und.nodak.edu/vwdocs/volc_images/img_vesuvius.html*

Pompeii: *www.archaeology.org/interactive/pompeii/history.html*

Mount St. Helens: *www.fs.fed.us/gpnf/volcanocams/msh* or *www.msnbc.msn.com/id/6092368*

Hawaiian volcano observatory: *http://hvo.wr.usgs.gov/maunaloa/current/main.html*

Volcano hazard mitigation: *www.geo.mtu.edu/volcanoes*

Types of volcanoes: *www.cotf.edu/ete/modules/volcanoes/vtypesvolcan1.html*

List of all volcanoes: *http://volcano.und.nodak.edu/vw.html*

Volcano simulator: *www.ees1.lanl.gov/Wohletz/Erupt.htm*

Volcano movies: *http://volcano.und.nodak.edu/vwdocs/movies/etna_mov.html* or *www.volcanovideo.com*

Lesson plan links

Living with volcanoes: *www.volcanolive.com/lesson2.html*

Understanding volcanoes: *http://school.discovery.com/lessonplans/programs/understanding-volcanoes*

Volcanoes!: *http://interactive2.usgs.gov/learningweb/teachers/volcanoes.htm*

In Flight, Online

ROBERT A. LUCKING, MERVYN J. WIGHTING, AND EDWIN P. CHRISTMANN

The concept of flight for human beings has always been closely tied to imagination. To fly like a bird requires a mind that also soars. Therefore, good teachers who want to teach the scientific principles of flight recognize that it is helpful to share stories of our search for the keys to flight. We would like to share some of these with you, using technology. The account of the work of the Wright brothers is one such a story. The idea of flight drew the brothers to the scientific articles in the Dayton (Ohio) Public Library in 1895 where Wilbur, age 28, and Orville, age 24, found their inspiration. Their reading led them to the work of a German inventor, Otto Lilienthal, who had conducted extensive glider experiments that reminded them of creations they had made from bamboo, cork, and rubber bands as teenagers.

During the next few years, the brothers built and experimented with several gliders and even built the first wind tunnel to test their work. They gathered detailed data on wind resistance, the efficiency of various wing designs, and propeller shapes that could be added to an engine. They recognized that they needed a strong headwind in order to experiment with different designs, and envisioned a place where they would encounter enough open space to launch their unpredictable craft. The location they eventually found was the 90-mile-long strand of sand jutting out into the Atlantic Ocean known as the Outer Banks of North Carolina. Near the small fishing village of Kitty Hawk in 1903, surrounded by the bright blue ocean and a broad expanse of sandy beaches, the brothers saw their dream realized. The details surrounding this story, lesson plans, and teaching resources are all available at the Smithsonian websites—*www.smithsonianeducation.org/educators/lesson_plans/wright.*

Plenty of other free print materials are also available at the Eisenhower National Clearinghouse—*www.enc.org*. Just type the name "flight" into the search engine. However, helping young people understand related principles of acceleration, drag, and aerodynamics involves more complex material, and the Eisenhower site also has a great book for middle school science students. Just type in "Flight Lab" to order the $21.80 package. Another great place to begin your exploration of materials on flight is the Smithsonian's more general site for teachers entitled How Things Fly—*www.smithsonianeducation.org/educators/lesson_plans/how_things_fly/lesson1_c.html.* Also, an excellent set of three lesson plans dealing with aviation and designed by PBS for grades 6–8 can be found at *www.pbs.org/kcet/chasingthesun/resources/resources_lessons.html.*

Middle school students love to experiment with hands-on material, so help them build their own planes by directing your class to a site that has a wide variety of paper plane designs at *www.paperairplanes.co.uk/index.html.* Yet another way to whet your students' appetite is to purchase the aviation adventure CD-ROM ($40). You can preview it at *www.gslis.utexas.edu/~kidnet/*

reviews/aviation.html and you will see that it will literally allow your students to have an adventure as they experience many different aspects of the world of flight. A commercial site on flight that works as another perfect launching point for students is *http://wings.avkids.com*. The NASA-sponsored K–8 Aeronautics Internet Textbook is available here. You will find lots of material for your students written both in English and in Spanish. Even more impressively, you'll find The Aeronautics Sign Language Dictionary. This site has more than you can imagine, especially if you are inclined to cross subject area boundaries. A special area exists for cross-curriculum work at *http://wings.avkids.com/Curriculums/Mythology/index.html.*

A great site, funded in part by NASA, for students to visit to learn about flight is *www.allstar.fiu.edu.* One section is designed specifically for teachers, and you'll find great material here broken out into three areas: history, principles of flight, and careers. In the portion on how airplanes fly, you'll find careful explanations and great visual aids in an article written by David Anderson of Fermi National Accelerator Laboratory. In the portion about parts of an airplane, students can learn of the structure of an aircraft including the fuselage, the landing gear, and the power source. The history of airplanes is provided, and it covers everything from balloons and airships to modern rocket development. The role of women in aviation is explored, and sections on minorities in the development of flight are included as well as a portion on the Tuskegee Airmen. Materials on careers involving flight are available, including the many different kinds of jobs at airports, in the military, and in aerospace. And the best news of all is that you can get all of the information described above for a voluntary contribution of $30 to Florida International University's Department of Mechanical and Materials Engineering.

Fortunately, many museums devote resources to helping teachers, and one website you'll want to have your students experience belongs to the Museum of Aviation at Robinson Air Force Base in Georgia: *www.museumofaviation.org/home.htm*. The opening visuals are so compelling that you will want to project them for your entire class, and the graphics dare you to click for more information. The photos are crisp and clear, the information about a wide variety of aircraft is first-rate, and the education section has a lot of great material that will appeal to middle school students. Another particularly useful site for materials related to flight comes from the San Diego Aerospace Museum: *www.aerospacemuseum.org*. The folks here have arranged information and activities around two themes, aviation and space flight. In both, they make materials available that fit within several disciplines; particularly useful to science teachers are the lesson plans on Newton's laws.

A commercial website you and students will enjoy is *www.howstuffworks.com/category-aviation.htm*. At this site you will find information about airports, airplanes, and airlines. You can even identify the formula for determining lift using the software FoilSim II, which allows you to investigate how an aircraft wing produces lift by changing the values of different factors that affect lift. There are several versions of FoilSim

II that require different levels of experience with the package, knowledge of aerodynamics, and computer technology. FoilSim was developed at the NASA Glenn Research Center in an effort to foster hands-on, inquiry-based learning in science and math. FoilSim is interactive simulation software that determines the airflow around various shapes of airfoils. The Airfoil View Panel is a simulated view of a wing being tested in a wind tunnel with air moving past it from left to right. Students change the position and shape of the wing by moving slider controls that vary the parameters of airspeed, altitude, angle of attack, thickness and curvature of the airfoil, and size of the wing area. The software displays plots of pressure or airspeed above and below the airfoil surface. A probe monitors air conditions (speed and pressure) at a particular point on or close to the surface of the airfoil. The software calculates the lift of the airfoils, allowing students to learn the factors that influence lift. The latest version of FoilSim (Version 1.5a) includes a stall model for the airfoil and a model of the Martian atmosphere for lift comparisons.

NASA Spacelink and other information providers across NASA are moving content onto the NASA home page at *www.nasa.gov*. This site is your starting point to explore NASA on the web, see showcased initiatives, and find breaking news. It is now the best place to find the type of content we have come to expect from Spacelink and is NASA's premier website. It includes an online user's manual that describes the various options available in the program and includes hyperlinks to pages in the Beginner's Guide to Aerodynamics describing the math and science of airfoils. More experienced users can select a version of the program that does not include these instructions and loads faster on your computer.

One of the best sites to explore concepts of flight is that developed by the NASA Glenn Research Center, *www.grc.nasa.gov/WWW/K-12/airplane/bga.html*. Here you'll be able to have students explore richly linked pages explaining how planes fly and the forces at work on planes. Their Learning Technologies Project includes great photos, lesson plans, and related websites; you'll have a hard time getting kids back from here. A full-scale replica of the Wright brothers' 1900 aircraft has been built in a joint effort between the Orono Middle School of Orono, Maine, and the NASA Glenn Research Center in Cleveland, Ohio. The aircraft has been used as a traveling exhibit, an educational tool, and eventually as a fixed exhibit at the Glenn Visitor Information Center.

An interactive simulation has been developed to predict the flight conditions necessary to fly this aircraft, *http://wright.grc.nasa.gov/replica/upredict.html*. Even more impressively, this site lists a variety of simulations that you can download for students, *http://wright.grc.nasa.gov/sim.htm*. In one such simulation students learn about the principles of aerodynamics by controlling the conditions of a big league baseball pitch, and in another, students can manipulate variables to design and test engines to see what factors make the engine most efficient. This software fosters hands-on inquiry-based learning in science and math by simulating jet engine tests and the effects that engines have on the speed and range of aircraft.

Internet resources

Smithsonian Lesson Plans
www.smithsonianeducation.org/educators/lesson_plans/wright

Eisenhower National Clearinghouse
www.enc.org

How Things Fly
www.smithsonianeducation.org/educators/lesson_plans/how_things_fly/lesson1_c.html

PBS Lesson Plans
www.pbs.org/kcet/chasingthesun/resources/resources_lessons.html

Paper Airplanes
www.paperairplanes.co.uk/index.html

Aviation CD-ROM
www.gslis.utexas.edu/~kidnet/reviews/aviation.html

The K–8 Aeronautics Internet Textbook
http://wings.avkids.com

Principles of Aeronautics
http://wings.avkids.com/Curriculums/Mythology/index.html

Aeronautics Learning Laboratory
www.allstar.fiu.edu

Museum of Aviation
www.museumofaviation.org/home.htm

San Diego Aerospace Museum
www.aerospacemuseum.org

How Stuff Works
www.howstuffworks.com/category-aviation.htm

NASA
www.nasa.gov

NASA Glenn Research Center
www.grc.nasa.gov/WWW/K-12/airplane/bga.html

Interactive Performance Predictions of Wright 1900 Aircraft Replica
http://wright.grc.nasa.gov/replica/upredict.html

Wright Flight Simulations
http://wright.grc.nasa.gov/sim.htm

Data Collection and Analysis Tools

EDWIN P. CHRISTMANN

The National Council of Teachers of Mathematics Standards expects that all students in grades 6 through 8 should be able to "formulate questions that can be addressed with data and collect, organize, and display relevant data to answer them" (NCTM 2000). Likewise, as proposed in the National Science Education Standards (NRC 1996), middle school students engaged in inquiry are to use mathematics and technologies to gather, analyze, and interpret data. Subsequently, objectives from mathematics and science lessons can be met efficiently, especially when technological tools such as graphing calculators, PDAs, and/or statistical software packages are used as tools for data analyses.

The following activity is a practical example of how middle school students can use technology to analyze data. In this illustration, students will be expected to design and conduct a scientific investigation based on data collected from height and arm span measurements (see Content Standard A: Science as Inquiry). Subsequently, after students make accurate measurements, they can design and execute investigations, interpret data, and form logical explanations about the results of the investigation. Keep in mind, however, that the technology applications applied here for data analysis can be used across a variety of topics. For example, the Principles and Standards for School Mathematics (NCTM 2000) gives several examples that show how middle school teachers can provide learning experiences that involve the analysis of data (see Figure 1).

Sample activity objectives

When you complete this lesson, you will be able to:

- formulate questions that can be addressed with data and collect, organize, and display relevant data;
- make and investigate mathematical conjectures;
- describe the shape and important features of a set of data and compare related data sets with an emphasis on how data are distributed;
- select and use appropriate statistical methodologies;
- compare different representations of the same data and evaluate how well each representation shows important aspects; and
- develop and evaluate inferences and predictions that are based on data.

Activity

During a life science unit on the human body, a class of sixth-grade students measured their height and arm spans (see Figure 2). This activity can be adapted for middle school students of differing grade and ability levels. For example, middle school teachers who work with gifted students will be able to have their students compute a variety of statistics—perhaps even a correlation coefficient or a t-test (Lehman and Christmann 1999). To be realistic, however, having students compute descriptive statistics (e.g., mean, median, mode, standard deviation, and so on), and organizing the descriptive statistics with a

Figure 1 Sample data analysis problems

- Compare the distance traveled by a paper airplane constructed using one paper clip, with the distance traveled by a plane that is built with two paper clips (here you can add a comparison among planes with three, four, five, or more paper clips). Which one travels farther when thrown indoors?
- Does a relationship exist between the length and width of warblers' eggs?
- Suppose you have a box containing 100 slips of paper numbered from 1 through 100. If you select one slip of paper at random, what is the probability that the number is a multiple of five? A multiple of eight? Is not a multiple of five? Is a multiple of both five and eight? (NCTM 2000)

Explore scale, powers of 10 at *www.scilinks.org.* Enter code SS30101.

statistical graph, such a scatter plot can be an excellent introduction to data analysis for most middle school students.

While students take the height and arm span measurements, the teacher can interface the TI-73 graphing calculator to an overhead projection panel to provide an audiovisual for the collection of data. Students can input their individual data quickly into the variables that the teacher has already designated in the data editor. The advantage in using a graphing calculator here instead of a spreadsheet program is the subsequent lesson's emphasis on data manipulation and interpretation. As an extension activity, students could use the math functions within a spreadsheet application to develop their own analyses. In addition, students can modify or update the data entry and see the modified results. The statistical graphics and tables can be placed into a finished research report.

Figure 2 Sample data

Subject	Height	Arm Span
1	142 cm	138 cm
2	148 cm	144 cm
3	152 cm	148 cm
4	150 cm	145 cm
5	141 cm	136 cm
6	142 cm	139 cm
7	149 cm	144 cm
8	151 cm	145 cm
9	147 cm	144 cm
10	152 cm	148 cm
11	150 cm	147 cm
12	152 cm	141 cm
13	148 cm	144 cm
14	152 cm	148 cm
15	144 cm	140 cm
16	148 cm	143 cm
17	150 cm	146 cm
18	138 cm	134 cm
19	145 cm	142 cm
20	142 cm	138 cm

After the height and arm span measurements are taken, the teacher can now lead a class discussion using a set of real-life data. Subsequently, the data can be organized in both table and plot formats in order to facilitate examination. For instance, a frequency table can be displayed and scanned for a minimum value, a maximum value, and a mode (if present). A box-and-whisker plot can be examined for outliers as well as for focusing on portions of the entire data set, namely quartiles (see Figure 3).

In addition to organizing data, data analysts generally summarize their data as well. Measures of central tendency (mean, median) and vari-

Figure 3 Box-and-whisker plot for the students' heights

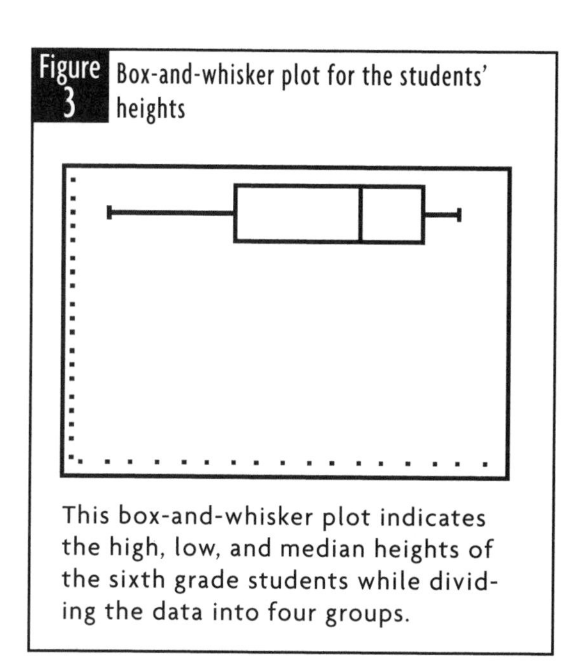

This box-and-whisker plot indicates the high, low, and median heights of the sixth grade students while dividing the data into four groups.

Figure 4 Descriptive statistics for students' heights

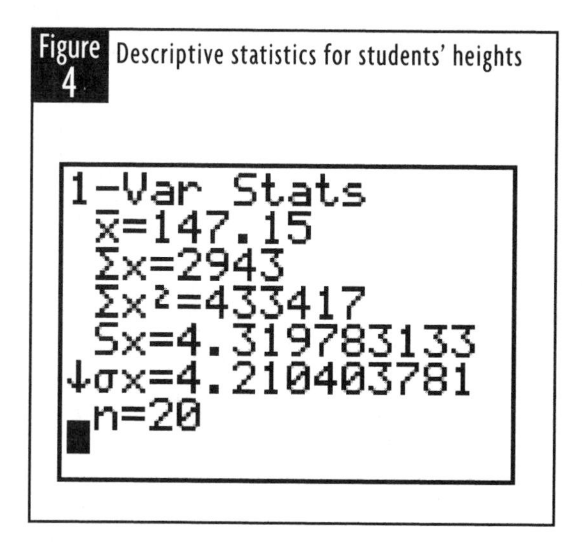

ability (range, standard deviation) are two of the descriptive statistics often used to "tell a story" about a set of data (see Figure 4). Again, computations like these can be performed on a single variable at a time or all variables simultaneously. Displaying all analyses or all raw data simultaneously can help students search for patterns among the computed statistics or raw data. For instance, students might detect that the lengths of height

Figure 5 Scatter plot of arm spans and heights

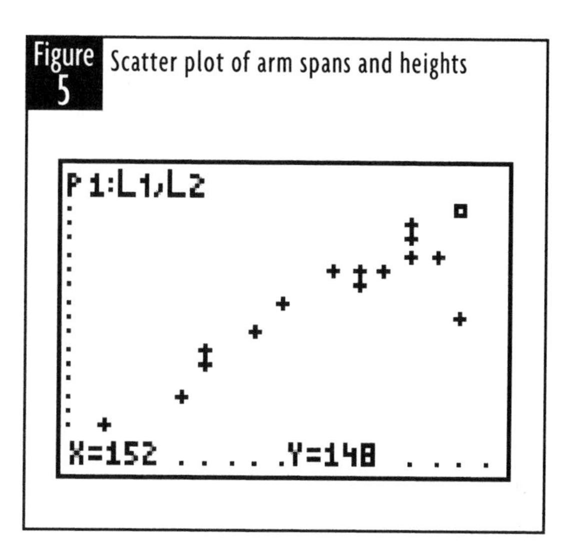

are longer than the lengths of arm span. In this example the arm span and height data appear to be similar. With the aid of the graphing calculator, the teacher can take such observations and have the class immediately examine them more quantitatively. For example, we can use the graphing calculator to draw a scatter plot showing the relationship between arm span and height (see Figure 5).

Besides analyzing entire data sets, scientists often also examine subsets of data. If each student inputs his/her gender at the time of data entry, the analyses previously described can now be performed on each subset of the data. Moreover, these follow-up analyses can be used to illustrate the point that sometimes conclusions that are drawn for a set of data may be largely influenced by a subset of those data. In later science classes, students will hear how chemists describe properties of elements but at times may subdivide the set of elements into the subsets of metals and nonmetals to examine their properties more closely.

Conclusion

Without a doubt, the incorporation of data analysis activities into middle school science instruction can improve the critical thinking skills of middle school science students. Simultaneously, middle school students will find themselves making connections and applications through the integration of mathematics, science, and real-world problems. As middle school teachers look for ways to integrate mathematics and science instruction, data explorations can help meet this ongoing objective.

References

Lehman, J. R. and E. P. Christmann.1999. Data explorations. *New Mexico Middle School Journal* 1(1), 26–28.

National Council of Teachers of Mathematics (NCTM). 2000. *Principles and standards for school mathematics.* Reston, VA: NCTM.

National Research Council (NRC). 1996. *National science education standards.* Washington, DC: National Academy Press.

Data Analysis and Probability Standard for Grades 6-8

The integration of data analysis into science instruction conforms to the tenets of the *Principles and Standards for School Mathematics* (NCTM 2000) as follows:

Instructional programs from pre-kindergarten through grade 12 should enable all students to:

- Formulate questions that can be addressed with data and collect, organize, and display relevant data to answer them.
- Select and use appropriate statistical methods to analyze data.
- Develop and evaluate inferences and predictions that are based on data.
- Understand and apply basic concepts of probability.

In grades 6 through 8 all students should:

- Formulate questions, design studies, and collect data about a characteristic shared by two populations or different characteristics within one population; select, create, and use appropriate graphical representations of data, including histograms, box plots, and scatter plots.
- Find, use, and interpret measures of center and spread, including mean, and interquartile range; discuss and understand the correspondence between data sets and their graphical representations, especially histograms, stem-and-leaf plots, box plots, and scatter plots.
- Use observations about differences between two or more samples to make conjectures about populations from which the samples were taken; make conjectures about possible relationships between two characteristics of a sample on the basis of scatter plots of the data and approximate lines of fit; use conjecture to formulate new questions and plan new studies to answer them.
- Understand and use appropriate terminology to describe complementary and mutually exclusive events; use proportionality and a basic understanding of probability to make and test conjectures about the results of experiments and simulations; compute probabilities of simple compound events, using such methods as organized lists, tree diagrams, and area models.

The Latest in Handheld Microscopes

MERVYN J. WIGHTING, ROBERT A. LUCKING, AND EDWIN P. CHRISTMANN

Around 1590, Zacharias Jansenn of Holland invented the microscope. Jansenn, an eyeglass maker by trade, experimented with lenses and discovered that things appeared closer with combinations of lenses. Over the past 400 years, several refinements to microscopes have occurred, making it possible to magnify objects between 200 and 1,500 times their actual size.

It was not until 1931 that Max Knott and Ernst Ruska invented the electron microscope, which allowed them to magnify objects up to 300,000 times their normal size. Today, the latest microscope technology can capture students' interests and imaginations with light, handheld imaging devices that can be connected to a classroom computer.

Handhelds:

ProScope

The ProScope (Figure 1) hooks directly to a USB port of a computer and is a snap to install and setup. The kit comes with a handheld camera, several lenses of varying magnification ranging from 1x to 200x, and a mount to view commercial slides. The bayonet-style lenses screw on easily, and all but the 1x magnification lens have a built-in light source. You and your students will be able to pick up the units and have them working in a very short time. Moreover, because of the simplicity of its operation, it is ideal for use as part of a student learning center or teacher demonstration.

These units include an amazing array of options that offer excellent demonstrations for the front of the classroom. Using this camera along with a computer hooked to a digital projector, you can show students details of life forms that are too small to see. The software has icons marked clearly so that you can choose live images, still shots, or short movies. Over time you will likely develop a collection of model images that integrate live shots of sample microorganisms gathered from outdoor lab activities. Still images can be saved in JPEG or BMP format, and movies in AVI. Captured images are displayed as thumbnails at the bottom of the software and can be enlarged.

The handheld nature of the scopes allows the user to touch the object or specimen to be viewed with the tip of the lens, moving the camera and specimen to get different views. Such use encourages interactivity, exploration, and observation of a specimen in real time, very often allowing students to see living microcosms that are difficult to observe. For example, the inside of a flower, feathers, bones, and cross sections of fruit can be examined by students. Also, try breaking a large sedimentary rock, such as sandstone or shale, into smaller pieces and have students explore the concept of sedimentation.

The ProScope has six white light-emitting diodes (LEDs) that surround its lens. The LEDs illuminate the object that you are looking at, and the image is displayed on your computer screen. It comes with a lens that magnifies objects up to 50 times (50x) their actual size, however, for $149 teachers can purchase a

Figure 1 Pro Scope

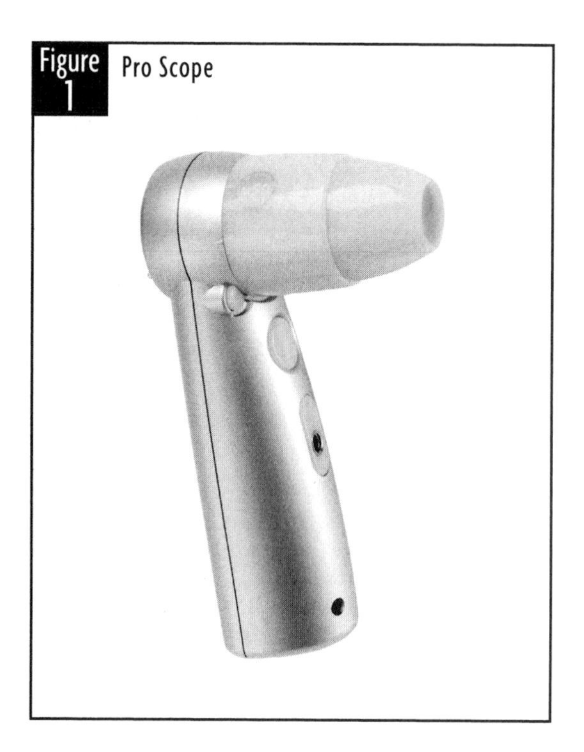

Figure 2 Scope on a Rope (SOAR)

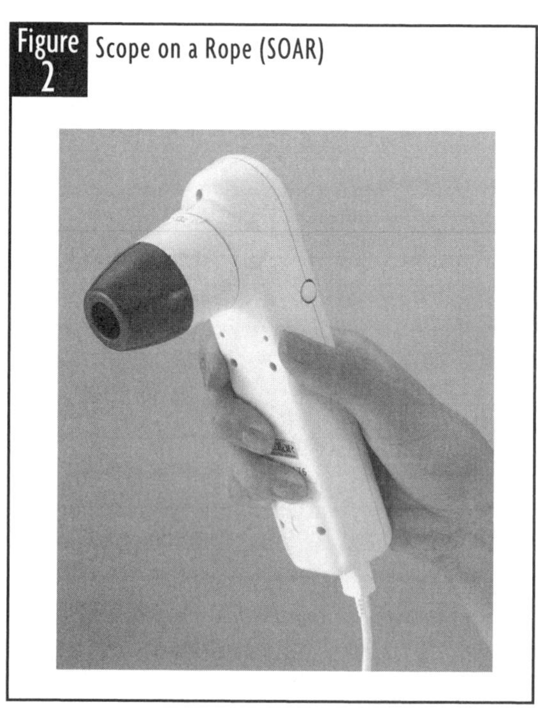

100x lens. For an even better view, a 200x lens is also available for $149. If you decide to use the accompanying stage set, which is a device that allows for hands-free viewing of prepared slides or water samples, you can view other objects as well, such as specimens in petri dishes. Educators can purchase a ProScope for $299. Details on ordering ProScope are available at *www.theproscope.com.*

Scope on a Rope

The second lightweight handheld imaging device is what has become known as Scope on a Rope (SOAR) (Figure 2). The SOAR, which is manufactured by ScalarScopes, was developed at Louisiana State University and was specifically designed to be an educational tool using Scalar's medical/industrial-quality camera and lenses. It is the original handheld device for education. The SOAR is unique among the handheld video microscopes because it is an analog camera with composite (RCA) video output. It will connect directly to any television set with a video input, any VCR, and any data projector for a live display. SOAR can also connect to a computer that is equipped with a third-party capture device to digitize the signal for display. The SOAR also has the most lens options with eight magnifications to choose from (1x up to 750x including a microscope adapter lens). Louisiana State University has produced resources to support the use of SOAR in the classroom, many of which are included in the basic kit. The equipment consists of a miniature, self-lighted video camera with interchangeable magnifying lenses. Each magnifying lens has a contact tip at the focal plane of the lens, so by simply touching the SOAR tip to the sample, a teacher or student automatically produces an in-focus image on the television screen. Little preparation of the specimen is necessary, which is a great advantage when compared to a conventional microscope.

The SOAR is very easy to use in a classroom setting and can perform advanced microscope functions as well. It is handheld and can be used by middle school students to provide instant in-focus images magnified on a regular classroom television. Its "real time" video camera provides normal-appearing motion for viewing living, moving specimens. The whole class can view a hair strand, the leg of a housefly, living organisms, a prepared slide, microfossils, or crystal structures, and you can be assured that all students are able to see—with a conventional microscope, this is not necessarily the case. Equally important, all students are able to view specimens at the same time and can respond to questions directed to the whole class. The teacher is freed from technical difficulties often encountered with school microscopes and video equipment, and does not need to assist students individually.

Students are able to see their own specimens shown on the classroom television—a medium that is often more appealing to them than other educational hardware! The teacher (or a student) can temporarily capture and "freeze" a frame on the monitor and discuss the detail of a particular specimen. The SOAR is very portable and is easy to hook up to a television. It requires very little space and is ready for use without any other accessories. It is probably best utilized if left set up in the classroom, readily available for spontaneous use by students.

The 30x and 200x objective lenses open new windows for observation and appreciation and will appeal to the inquisitive mind of the middle school student. This SOAR also has a regular or infinity lens that can be interchanged with the magnifying lenses. This lens allows the handset to be used like a regular video camera, with the ability to focus at infinity and also up close.

Scientific images produced on the television using a SOAR may be videotaped with a VCR in the normal way. This will allow groups of students to share the results of their experiments with the whole class, or a class project could be shown to other students or groups of parents. Science teachers can use the SOAR to make films for the whole class that might form an integral part of a lesson. A video-printer can be plugged into the SOAR to produce instant photo-quality prints that a student can use as part of a project or to record in a portfolio. Also, a computer capable of video-capture can be used with the unit to obtain computer printouts or prepare digital files of television images.

Middle school teachers of subjects other than science may also find the SOAR useful. For example, it can be used to stimulate creative writing by showing the class the magnified image of a drop of dew on a rose petal, the "huge" hairy leg of a spider, or the enhanced picture of a mealworm. Examination of minute structures using the SOAR can lead to a discussion of mathematical concepts and geometric shapes. A field trip to a local beach or pond could be conducted to collect specimens (collecting jars are included in the basic kit), and the trip itself could be used as a stimulant for writing across the curriculum.

A CD guide for teachers comes with the SOAR that incorporates a large number of recorded images of magnified specimens. Also included are sample lesson plans that are designed for middle school students. Examples include a lesson to compare different soil samples to determine which soil characteristics are best for supporting life; a study of absorbent and non-absorbent materials readily available in the classroom (such as paper towel, sponge, plastic and rubber eraser); and a project to investigate the relationship between a newly hatched insect and the mature larva that produced it. These lesson plans are comprehensive, well documented, and relate to a number of middle school science standards. Scope on a Rope is available as an Education Bundle, with prices ranging from about $1,000 to $3,000. For more information and sample activities, visit *www.scopeonarope.lsu.edu* and *www.scalarscopes.com*.

Figure 3 Swiftcam system by Swift Instruments, Inc.

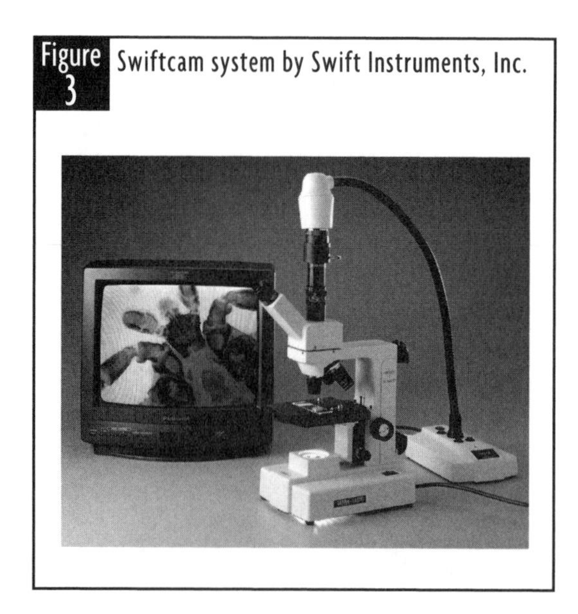

Figure 4 Pupil CAM

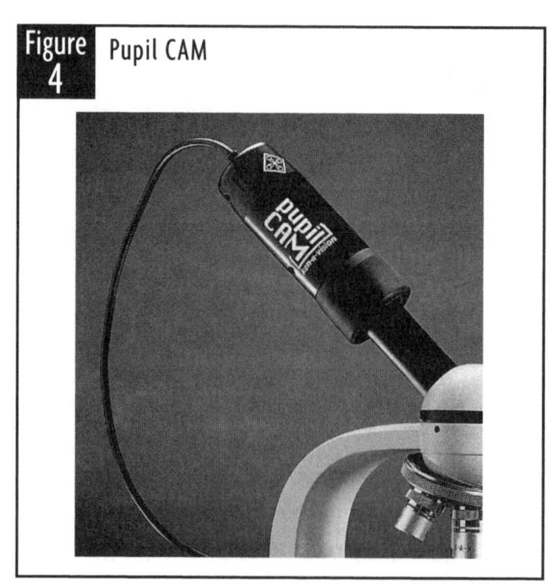

Other options:

Video Flex and Swiftcam

These two imaging systems, from Ken-a-Vision and Swift Microscopes respectively, are attached to a flexible stand that allows the operator to position the lens quickly and easily (Figure 3). They can be used by themselves or attached to the eyepiece of a standard microscope to increase the magnification of your images. The images can be displayed on a TV or computer, or shown onscreen with an LCD projector. You can capture still images on your computer or save video clips through a VCR. Systems range in price from about $600 to $1,300. Visit *www.kenavision.com* and *www.swiftmicroscope.com* for more information.

Pupil CAM

The Pupil CAM digital microscope camera connects to the eyepiece of most microscopes (Figure 4). Cabling is preattached to the camera that allows you to view the images on either a TV or computer; however, an adapter is available to convert the video connection to a computer connection. Software is provided to help you store, share, and manipulate your images on your computer. An optional 10x built-in eyepiece is available to boost the magnification of your microscope. Each unit costs $290, but classroom sets are available at a discount. For more information, visit *www.ken-a-vision.com/pupilcam.asp.*

Conclusion

Undoubtedly, the potential for handheld microscopes is boundless. However, if you choose to adopt handheld microscopes, the scientific applications are not much different from those of traditional microscopes. Therefore, make purchasing decisions based on learning objectives, annual budgets, curriculum requirements, and the compatibility of the handheld microscope with the equipment that is currently in place. The ideas presented here will update your knowledge of the technologies available for examining the structure and function of living things in the middle school science classroom. Integrating the latest technologies into your science classroom allows students to make real-world connections between scientific data and their daily lives.

Reference

National Research Council (NRC). 1996. *National science education standards.* Washington, DC: National Academy Press.

National Standards

The use of handheld microscopes in the science classroom provides teachers and students with a unique opportunity to do demonstrations and laboratories that conform to the specifications of the Standards (NRC 1996). Clearly, the use of laboratory activities that use handheld microscopes is a method for science teachers to integrate the abilities that are proposed in the Life Science Content Standard C: Grades 5–8. Students should have the fine motor skills to work with a light microscope and interpret accurately what they see; structure and function in living systems. Students should incorporate the use of computers and conceptual models; and design and conduct scientific investigations.

Aqua Analysis

JUSTIN SICKLES, EDWIN P. CHRISTMANN, AND JEFFREY LEHMAN

The relationship of science to personal health becomes a relevant issue to middle school students as they grow aware of their physical development and the importance of good health. Nutrition plays an important role in physical health, and proper nutrition includes the intake of adequate amounts of drinkable water. Students can use the following internet-based activity to examine the rapidly expanding commodity of bottled water.

Bottled water boom

With the increasing availability of internet access in schools, middle school teachers have access to a powerful and expansive tool to investigate consumer products. Bottled water has become big business—in 1999, Canadians alone consumed 703 million liters of bottled water (Nichols 2000). The popularity of bottled water is due, in part, to its perception as a pure source of water. Nearly all water bottlers and bottled water suppliers sponsor homepages (such as *www.water.com* and *www.bottledwaterweb.com*) that provide information about their products and the purification processes used during production.

Bottled water labels may provide some product content information, although few companies, if any, list this information on their websites. However, results of independent research on nearly all brands of bottled water are posted on internet sites such as the Drinking Water Research Foundation (*www.dwrf.info*), and the National Sanitation Foundation (*www.nsf.org*). Using the internet, you can provide students with a vast source of data regarding bottled water purity.

Internet investigations

A few days before you conduct the activity, ask students to write down as many bottled water brands as they can find. These brands may include distilled, seltzer, mineral, and spring water. Create a class list from the results and use it to assign students to work in cooperative research groups.

On the day of research, which may be extended beyond one class period, students use the internet to obtain data for specific values of mineral content, such as calcium, magnesium, and sodium—and contaminants, such as arsenic, chloroform, and nitrates. Students may want to examine the history of water safety standards, such as the Safe Drinking Water Act, passed by Congress in 1974. They also may want to compare the purity standards set by the Food and Drug Administration (FDA) and those of the Environmental Protection Agency (EPA). Find these standards online at *http://bottledwater.org/public/BWFactsHome_main.htm* and *www.epa.gov/safewater/standards.html,* respectively.

Diving into the data

Students may organize this information using spreadsheet software. Figure 1 shows a template from which students can start. Students can expand this simple spreadsheet to include more categories and more brands as they gather

Bottled water mineral and contaminant content

Brand	Calcium (mg/L)	Magnesium (mg/L)	Arsenic (ppb)	Chloroform (ppb)
Arrowhead	20	5	3.2	0
Black Mountain	25	1	0	1.4
Crystal Geyser	0	6	11	0
Deer Park	1	1	0	0
Evian	78	24	2	0
Naya	38	20	0	0

data. To integrate mathematics, ask students to calculate statistics for each mineral and contaminant: mean, minimum, maximum, sum, standard deviation, variance, standard error of the mean, and range, as prescribed by the National Council of Teachers of Mathematics' (NCTM) National Mathematics Standards (NCTM 1989).

Students begin by entering the data into the spreadsheet. Students can then use the spreadsheet functions to report average content of specific minerals or contaminants for all brands investigated; minimum and maximum levels obtained for each category; and any patterns or anomalies they notice. You may ask, "Which brands contain the most contaminants?" "Which appeared to be the purest?" Students also may use the spreadsheet software to create graphs or charts. (See Figure 2 for an example.) These visual representations of the data may facilitate or enhance students' understanding of the data. You may suggest that students create graphs to show which brand contained the highest level of nitrates, or which contained the lowest amount of calcium.

You can expand the activity by asking students to compare contamination data obtained for brands that have multiple sources or bottling sites. Find this information at sites such as that of the National Resources Defense Council (*www.nrdc.org/water/drinking/nbw.asp*). You may ask students questions such as, "For a certain brand, why may contaminants be present at one location and not another?" "Is it the water source or the purification method that results in the contamination?" "Is there a way to know for sure?" "Could something else cause the appearance of contaminants?" "Can you be assured of getting uncontaminated water every time you purchase bottled water?"

As students gain more experience in exploring the internet and become more familiar with the spreadsheet software, they may return to investigating the standards of bottled water purity set forth by the FDA and the EPA. Students may compare and contrast the guidelines that each of the agencies establish, and may create another spreadsheet to compare the agencies' values. You may ask students, "Why do the values vary for certain items?" "Are one agency's standards better than another's?" "What are some concerns in establishing values for contaminants?" "What are some possible effects of these standards on the population?" "What impact do these standards have on bottled water producers?"

As a follow-up, consider inviting a community resource person from the water authority that provides drinking water to the students' school.

Figure 2 Calcium levels in bottled water

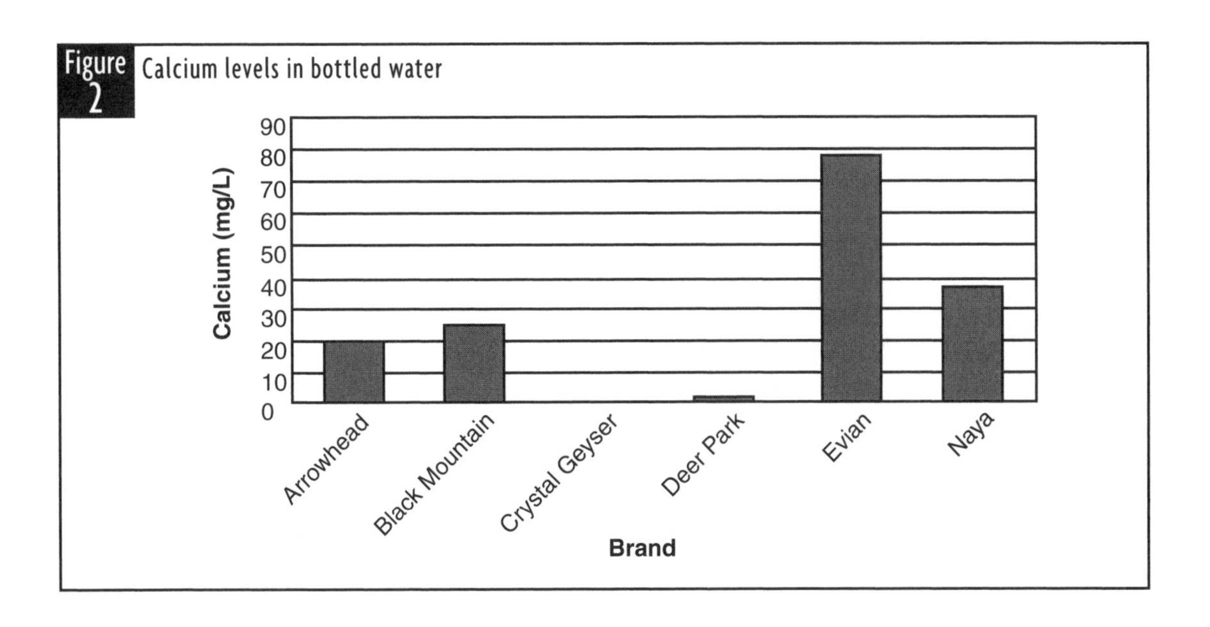

Refreshing results

This activity integrates many ideas present in science education standards for middle school students. It incorporates mathematics and the use of appropriate tools and techniques to gather, analyze, and interpret data, all advocated by the National Science Education Standards. By providing students with a familiar (and increasingly popular) topic, they will learn to use and incorporate the internet and spreadsheet software. Using common products, students are able to make real-world connections in their science classrooms and at the same time become more aware of environmental issues as current and future consumers.

Both the National Science Education Standards and state-level standards advocate a changing emphasis in science education. These standards place a greater emphasis on inquiry, technology, and science in personal and social perspectives (NRC 1996).

References

Bybee, R. W., and A. B. Champagne. 1995. National science education standards: An achievable challenge for science teachers. *The Science Teacher* 62(1): 40–45.

National Research Council (NRC). 1996. *National science education standards.* Washington, DC: National Academy Press.

National Council of Teachers of Mathematics (NCTM). 1989. *Curriculum and evaluation standards for school mathematics.* Reston, VA: Author.

Nichols, M. 2000. Is bottled better? Canadians drink millions of litres—so safety is paramount. *Macleans* (June): 26.

Fast-Food Fact Finding

EDWIN P. CHRISTMANN, JILL KONTON, AND JEFFREY LEHMAN

During the middle school years, students begin to broaden their basic understanding of health issues and learn about the risks and benefits associated with their dietary decisions. One of the goals of the Standards at this age level is to develop a scientific understanding of health. This activity takes advantage of your students' knowledge of popular culture and the internet by examining a staple of most students' lives—fast food (Cothron, Giese, and Rezba 1993).

Fast-food frenzy

The popularity of the internet has made accessible numerous consumer information sites. Nearly every major fast-food chain has established its own homepage (for example, *www.mcdonalds.com, www.burgerking.com,* and *www.wendys.com)* providing information on menus, promotions, company history, and nutritional analysis of its food. Using this information, students can begin to compare and contrast the nutritional value of menu items from different restaurant chains.

Specific categories students can examine include total calories, calories from fat, total fat, cholesterol, sodium, carbohydrates, and dietary fiber. You may ask how a typical fast-food meal compares with the recommended daily allowances set by the FDA (Saltos 1993). The allowances suggest that average men need around 2,700 calories per day and women need around 2,000 calories per day to maintain a desirable weight. Middle school students typically require more.

The American Heart Association suggests limiting fat intake to less than 30 percent of daily caloric intake: no more than 50–80 grams of total fat and 300 milligrams of cholesterol per day. Sodium intake should fall between 1,100 and 3,300 milligrams, or 1/2–1 1/2 teaspoons per day.

Good nutrition is extremely important for middle school students because it is during these few years that they may acquire 15 percent of their adult height, 50 percent of their adult weight, and a large part of their maximum bone mass (Chicoye, Jacobsen, and Landry 1997; Sinclair 1978). (This is based on an approximation of human development between the years of 13 and 15.) This project alerts students to the failure of many fast foods to meet the recommended dietary standards for fiber and other important nutrients while exceeding those for fat, sodium, and other unwanted elements.

Spreadsheet savvy

Because "data handling is at the heart of science," spreadsheets can provide a valuable tool to extend student investigations (Pogge and Lunetta 1987). You can create templates in advance that highlight aspects for comparing fast-food products (see Figure 1). Students could expand their investigation by including as many fast-food chains as possible (to provide an accurate overview of fast food).

After entering the initial data into a spreadsheet, students can add additional information and manipulate the data to study the relationships among different categories. For instance,

Fast food nutrition data for hamburgers

Restaurant	Calories	Fat (g)	Cholesterol (mg)	Sodium (mg)
McDonald's	270	9	30	530
Burger King	260	10	30	500
Wendy's	270	10	30	560
Hardee's	260	9	20	460
Dairy Queen	290	12	45	630
White Castle	161	8	0	266

Energy calculations

Steps
- Multiply the number of grams per serving by percent fat to obtain the number of grams of fat per serving.
- Multiply the number of grams of fat by 9 calories/gram of fat.
- Divide the number of calories from fat by the number of calories in a serving to determine the percentage of fat per serving.

Example
- Lean hamburger
- Serving size: 100 grams
- Percent fat by mass: 5% fat
- Calories per serving: 179

Calculations (per serving)
- 100 grams x 5% fat by mass = 5 grams of fat per serving.
- 5 grams of fat x 9 calories/gram of fat = 45 calories.
- (45 calories/179 calories) x 100% = 25%

an important concern with most fast food is the percentage of calories from fat. (See Figure 2 for how to determine this value.) Students can simply enter the formula into a spreadsheet to calculate this percentage automatically.

Students can also make use of spreadsheet functions such as SUM (for the total calories of a meal) and AVERAGE (for the average amount of sodium in fast-food hamburgers). Graphing and plotting data can further enhance the students' understanding of the relationships among the data (see Figure 3). To spark ideas on what students can graph, you can suggest investigating which chain makes the saltiest french fries or how the amount of fat relates to the calories in a burger.

After organizing the data collected, you might ask students higher-order thinking questions, such as "Why are some burgers lower in fat than others?" "How might the way that this food is prepared affect these categories?" "Is a flame-broiled burger really better for you than a fried burger?" "What other categories for describing burgers can be added to the spreadsheet?" "How do condiments affect the burgers' nutritional content?" and "Which burgers give you more nutritional value for your money?"

As students become more involved with exploring the internet for raw data and company information, they may want to investigate related issues and sites. For instance, independent sites such as *www.olen.com/food/book.html* provide consumers with quick nutritional facts on popular fast-food restaurants as well as healthier alternatives on fast-food menus. You may wish to prompt further student investigations by asking questions such as "What was the first fast-food restaurant and when did it start?" "Why is fast food so popular in today's society?" or "How does the media affect our decisions on what and where we eat?" This type of questioning gets students to ponder the relationships among technology, society, dietary decisions, and the risks and benefits that accompany a fast-paced lifestyle.

Fun food findings

This project integrates a number of research standards outlined in the National Science Education Standards (NRC 1996). For example, this in-depth analysis of fast food focuses on Life Science Standard C, regulation and behavior, and also emphasizes personal health, measurement, and the statistical analysis of data. Given a popular and familiar topic to explore, students will embrace the internet and spreadsheet software as tools for learning. Integrating projects like this into the science curriculum allows students to make real-world connections between scientific data and their everyday lives.

Figure 3 Energy calculations

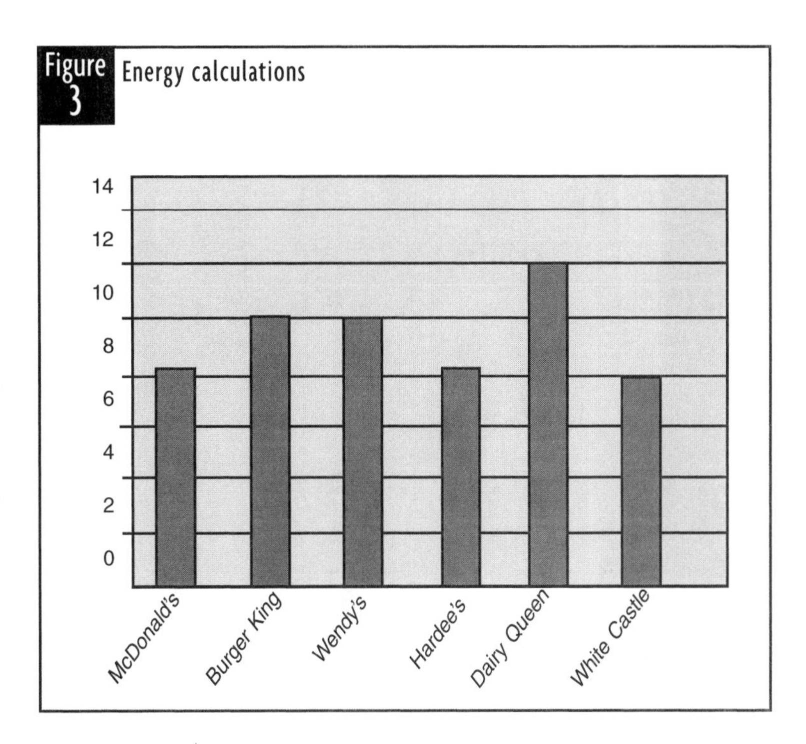

References

Chicoye, L., M. Jacobsen, and G. Landry. 1997. "Getting teens fit and well-nourished: Shaping the future." *Patient Care* 31(12): 72–74.

Cothron, J., R. Giese, and R. Rezba. 1993. *Students and research: Practical strategies for science classrooms and competitions.* Dubuque, IA: Kendall/Hunt Publishing Company.

National Research Council (NRC). 1996. *National science education standards.* Washington, DC: National Academy Press.

Pogge, A. F., and V. N. Lunetta. 1987. Spreadsheets answer "What if . . . ?" *The Science Teacher* 54(8): 46–49.

Saltos, E. 1993. The food pyramid-food label connection. *FDA Consumer* (May): 17–21.

Sinclair, D. 1978. *Human growth after birth.* London: Oxford University Press.

Testing the pH of Soft Drinks

EDWIN P. CHRISTMANN AND ADAM J. HOLY

In physical science, students are often asked to identify a substance as an acid or a base through a litmus test. A litmus strip turns faded violet when dipped in a neutral liquid, pink when dipped in an acidic liquid, and blue when dipped in a basic liquid. Although a litmus test is useful to identify acids and bases, it does not give an exact measurement of the pH of a substance. To obtain a numerical value for pH, students can use a probe connected to a graphing calculator or computer as part of a hands-on investigation of something all students can identify with—soft drinks.

In this article, I describe how to use a TI-73/83/84 graphing calculator and Vernier's LabPro/CBL2 probe system to take pH readings. This is not an endorsement of these products, but simply my attempt to give you an idea of what is involved in using this technology in the classroom. This activity can be accomplished using a variety of probeware connected to a laptop, desktop, or PDA. The activity can easily be adapted to whatever hardware you have on hand.

What is pH?

pH is a unit of measure that is used to express the degree of acidity of a substance. The pH notation is an index of hydrogen's chemical activity or its potential in a solution. The pH scale goes from 0 to 14, with a pH of 7 considered neutral. A substance with a pH below 7 is considered acidic, and one with a pH above 7 is considered basic (alkaline).

The pH of a solution can be measured with a pH meter with an electrode system whose voltage output is proportional to the active acid (H_3O^+) concentration in solution. Distilled water has a pH of about 7. Tap water varies between a pH of 6.8 and 7.2 and is generally slightly alkaline in hard water areas (Joyce 2003).

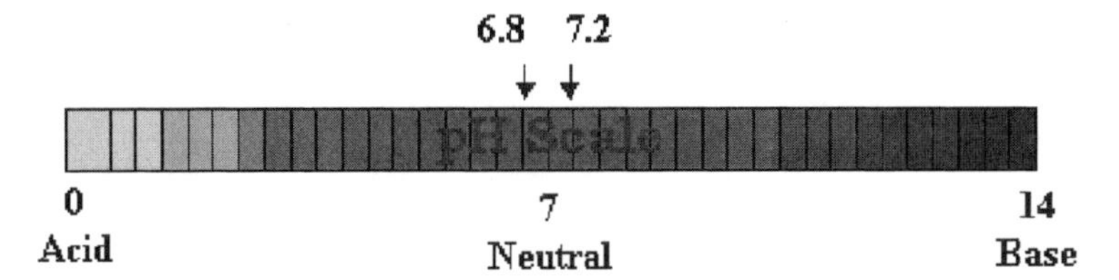

Testing the pH of soft drinks

In this activity, you will predict the pH of several soft drinks. You will then use a probe and graphing calculator to determine the actual pH of each soft drink and compare the results to your predictions. You will then come up with your own question to investigate about the pH of soft drinks.

Materials

- 100-mL samples of various carbonated drinks, such as Coke, Pepsi, and Mountain Dew
- safety glasses
- tissues or paper towels
- 250 mL beakers
- 500 mL distilled water
- buffer solutions of pH 7 and pH 10 (available from science suppliers)

- CBL/CBL2
- TI-83 plus silver edition calculator
- DataMate program for TI-84
- TI connectivity cable
- TI-graph link cable
- unit-to-unit link cables
- Vernier LabPro pH sensor

Testing procedure

1. Plug the pH Sensor into Channel 1 of the LabPro or CBL 2 interface. Use the link cable to connect the TI graphing calculator to the interface. Firmly press in the cable ends.
2. Turn on the calculator and start the Data-Mate program. Press CLEAR to reset the program.
3. Set up the calculator and interface for the pH Sensor.
    a. Select SETUP from the main screen.
    b. If the calculator displays pH in CH 1, proceed directly to Step 4. If it does not, continue with this step to set up your sensor manually.
    c. Press ENTER to select CH 1.
    d. Select pH from the SELECT SENSOR menu.
4. Set up the calibration for the pH Sensor.
    - If your instructor directs you to use the stored calibration, proceed directly to step 5.
    - If your instructor directs you to manually enter the calibration values, select CALIBRATE, then MANUAL ENTRY. Enter the slope and the intercept values for the pH calibration, select OK, then proceed to Step 5.
    - If your instructor directs you to perform a new calibration for the pH Sensor, follow this procedure:

First calibration point

a. Select CALIBRATE, then CALIBRATE NOW.
b. Remove the sensor from the bottle by loosening the lid, then rinse the sensor with distilled water.
c. Place the sensor tip into pH-7 buffer. Wait for voltage to stabilize, then press ENTER.
d. Enter "7" (the pH value of the buffer) on the calculator.
e. Rinse the pH Sensor with distilled water and place it in the pH-10 buffer solution.
f. Wait for the voltage to stabilize, then press ENTER.
g. Enter "10" (the pH value of the buffer) on the calculator.
h. Select OK to return to the setup screen.

5. Set up the data-collection mode.
    a. To select MODE, press % once and press ENTER
    b. Select SINGLE POINT from the SELECT MODE menu.
    c. Select OK to return to the main screen.
6. Collect pH data.
    a. Remove the pH Sensor from the storage bottle. Rinse the tip of the sensor thoroughly with the distilled water.
    b. Submerge the sensor tip to a depth of 3–4 cm into the beaker containing your first carbonated beverage sample.
    c. When the readings stabilize, select START to begin sampling. **Important:** Leave the probe tip submerged while data are being collected for 10 seconds.
    d. After 10 seconds, the pH value will appear on the calculator screen. Record this value.
    e. Press ENTER to return to the main screen.
    f. Select START to repeat the measurement (round to the nearest 0.01-pH unit).
    g. Press ENTER to return to the main screen.
    h. Rinse the sensor with distilled water and

return it to the storage bottle when you have finished collecting your data.

7. Record pH reading.

	Predicted pH	Actual pH
Sample 1		
Sample 2		
Sample 3		
Sample 4		

Exploring pH

Design an investigation to explore a question about pH. For example, you could aks

- Does the pH of soda change when it goes flat?
- Does temperature affect pH?
- Is the color of a beverage an indicator of its pH?
- How does the pH of diet sodas compare with regular sodas?
- How does the pH of sodas compare with juices and sport drinks?

National Standards

Having middle school science students explore acids and bases is commensurate with Content Standard B of the National Science Education Standards in that middle school science students should explore the different characteristic properties of substances (NRC 1996).

References

Joyce, R. J. 2003. *pH simplified*. Available online at *www.luminet.net/~wenonah/hydro/ph.htm* (accessed April 4, 2005).

National Research Council (NRC). 1996. *National science education standards*. Washington, DC: National Academy Press.

Note

According to company information specialists, Coca-Cola Classic has a pH of approximately 2.3 and Pepsi has a pH in the range of 2.5–3.5.

Graphing Calculators

EDWIN P. CHRISTMANN

The middle school science curriculum should include explorations of statistics in real-world situations that require students to collect, organize, and represent sets of data. A graphing calculator is the perfect tool to help you achieve this goal. My personal favorite, the TI-73 graphing calculator from Texas Instruments, has built-in mathematical functions for middle school students, but less sophisticated calculators will work just as well for data analysis activities.

Heightened calculations

The following activity represents a practical example of how a middle school teacher can use the TI-73 calculator to analyze students' height measurements during a life science unit based on the question: "How do the heights of boys and girls in this class compare to boys and girls throughout the United States?" Correspondingly, the activity fulfills the tenets of Content Standard A: Science as Inquiry of the National Science Education Standards (NRC 1996).

Data collection

This activity begins with a demonstration of how data are entered into the graphing calculator. With the TI-73, this can be done by interfacing the calculator with an overhead projector so that the entire class can follow along at the same time. Using the TI-73's keys, data entry should only take a few minutes. Following the teacher's demonstration, students can create an appropriate table prior to the activity similar to Figure 1 that is stored in the calculator's list editor.

Figure 1 Sample TI-73 data entry screen

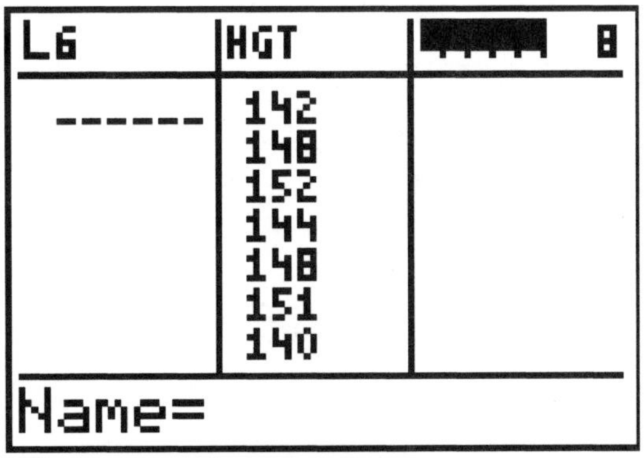

L6	HGT	8
------	142	
	148	
	152	
	144	
	148	
	151	
	140	

Name=

However, prior to data entry, the next step is to have students measure the height of everyone in the class. These data are then entered into a table on the graphing calculator. Figure 1 is an example of a table created for this activity. It displays the height in centimeters of seven sixth-grade boys.

Data analysis

After data is entered, the graphing calculator does all of the other calculations with the touch of a few buttons. Figure 2 shows a variety of one-variable statistics that can be calculated with the TI-73. For advanced students, the mean and standard deviation information could lead to a discussion

Figure 2 One-variable statistics

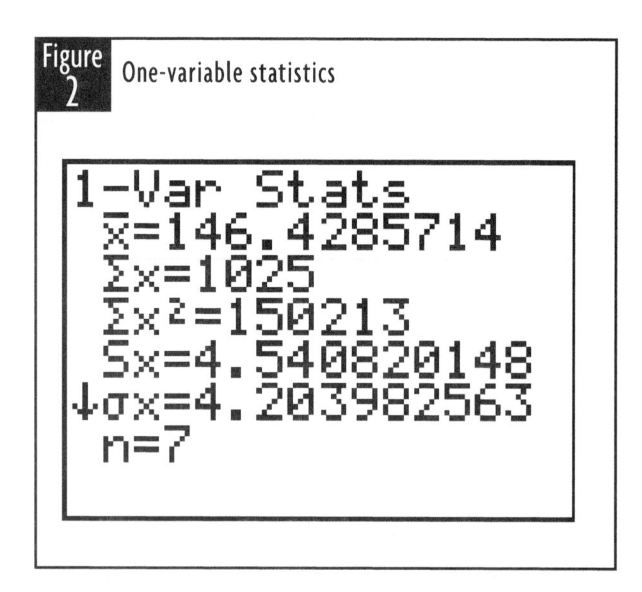

Figure 4 Box-and-whisker plot

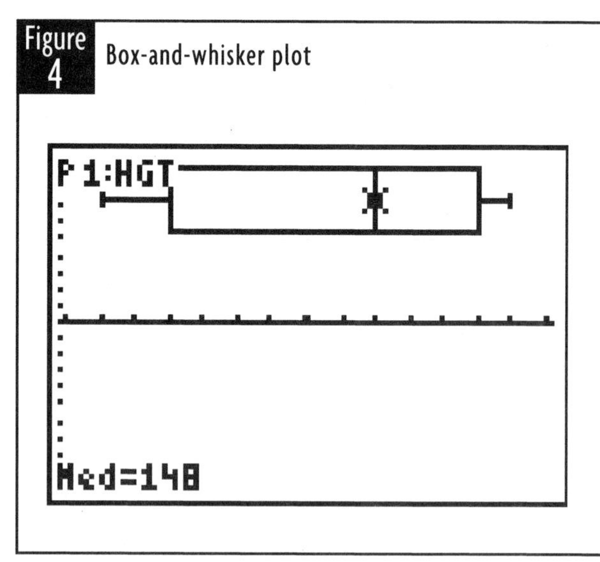

Figure 3 Median and mode computations

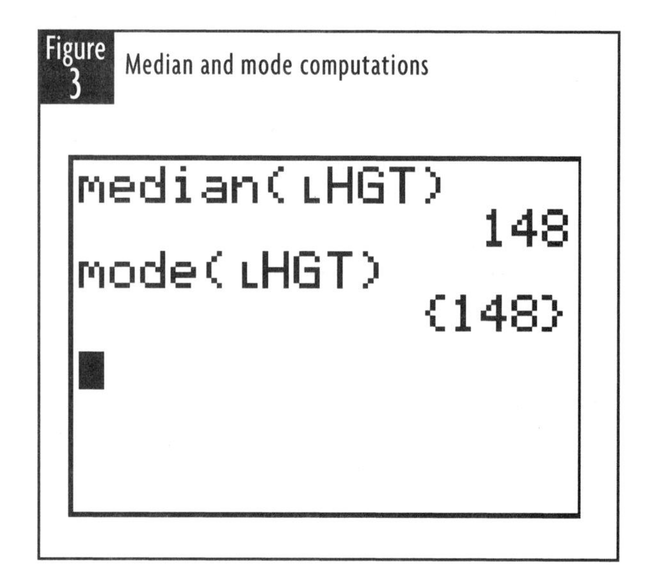

of inferential statistics and scientific inquiry. The TI-73 also allows you to calculate two measures of central tendency—the median and the mode (see Figure 3). This is a very convenient, timesaving feature because students do not have to form an array and frequency table to calculate the median and mode.

With the results of the data-analysis, students can report the mean heights for boys and girls, their minimum and maximum heights, the range in heights, and any noticeable patterns. For example, students can compare measures of central tendency between boys and girls. Why is there less variation among boys? How does the height of the tallest boy compare to the height of the tallest girl? Students can suggest possible explorations for these and other patterns they uncover.

The TI-73 can also be used to introduce a different type of graph—the box-and-whisker plot (see Figure 4). The box contains the central 50% of the distribution, from the lower quartile to the upper quartile. The median is marked by a line drawn within the box. The lines extending from the box are called "whiskers." In the plot, the data are divided into four quartiles: the left whisker, the right whisker, the left side of the box, and the right side of the box. Students should try to explain how the data are distributed and what each quartile represents. Data for girls can also be graphed so students can compare the two plots.

After the initial data analysis, students can go back and answer the initial question, (i.e., "How do the heights of boys and girls in this class compare to boys and girls throughout the United States?"). Using the internet, students can find answers from several internet sites, such as the

CDC's Vital and Health Statistics site at *www.cdc.gov/nchs/about/major/nhanes/growthcharts/clinical_charts.htm*, which provides students with the national height averages for all children and adolescents up to age 20 in the U.S.

Finally, if a computer station is available in the classroom, teachers can load TI-Graph Link software onto their computers for free; which can import data analysis results from the calculator to word processing software such as MS-Word. This allows students to create slick and professional looking lab reports.

Summary

Although the TI-73 calculator was the tool used for data analysis in this activity, you could use a less sophisticated calculator or a more complex computer-based statistical software package to do the calculations. However, a major advantage of using a TI-73 graphing calculator is that it's intended for data manipulation and interpretation that is designed for use in the middle school classroom. Moreover, calculators are more portable than computers and approximately 20 students can be equipped with a graphing calculator for the price of a single microcomputer workstation.

Standards

According to the NCTM's Data Analysis and Probability Standard (NCTM 1998), the mathematics and science curricula for students in grades 6–8 should enable all students to "Formulate questions that can be addressed with data and collect, organize, and display relevant data to answer them." Moreover, the NCTM suggests that instructional programs should enable students to select and use appropriate statistical methods for data analysis. Likewise, as proposed in the National Science Education Standards (NRC 1996), middle school students engaged in inquiry are to use mathematics and technology to gather, analyze, and interpret data. Simultaneously, data explorations meeting curricular objectives can be enhanced when technological tools are used to process data.

References

National Council of Teachers of Mathematics (NCTM). 1998. NCTM principles and standards for school mathematics, electronic version: Discussion draft. Reston, VA: Author.

National Research Council (NRC). 1996. National science education standards. Washington, DC: National Academy Press.

Personal Digital Assistants (PDAs)

EDWIN P. CHRISTMANN

The personal digital assistant (PDA) is rapidly becoming one of the most popular technological tools available for data organization. Not to be mistaken with the pocket PC, the PDA is a small hand-held organizer that uses the Palm operating system and a stylus to enter and extrapolate data. Although Palm, Inc. pioneered the PDA, several clones are now available from companies such as Xircom, Royal, and Visor, with prices ranging from $150 to about $450 for high-end models. Therefore, a $3,000 investment could equip a science laboratory with 15 PDAs and plenty of software applications.

So how can PDAs be used in educational settings? According to Palm Inc., "...a student, teacher, or administrator can access the internet wirelessly, take class notes, calculate, sketch ideas, collect data, access resources, manage activities and courses, and instantly beam information to others. Palm handhelds offer personal and pervasive access to computing power that can stimulate active, inquiry-based learning and lead to increased productivity and effectiveness. A Palm handheld, weighing in at about five ounces, has enough functionality and storage to replace much of the 20 pounds of learning materials—reference and other books, study guides and worksheets, organizers, and graphing calculators. And accessories such [as] MP3 players and cameras can be added" (Palm, Inc. 2001).

Based on the reported capabilities, there is little doubt that the PDA has the potential to offer teachers an additional tool to support instruction. Moreover, the National Educational Technology Standards (NETS) for Students suggest that students should have access to various aspects of technology (ISTE 1998). Table 1 lists the basic tenets of the NETS, while reinforcing the premise that the incorporation of PDAs into contemporary science instruction is germane.

Obviously, teachers should take these standards into consideration when using any classroom technology; however, the point here is to give teachers a reference, and to illustrate how the reported software applications available for the PDA align with the NETS. Clearly, it is very important that teachers take the national science and mathematics standards into consideration when planning classroom activities as well.

The samples of PDA software in Figures 1 and 2 should show middle school teachers how they can use the PDA for labs, references, calculations, and as a classroom management tool in their science classrooms. Likewise, it is important to note that teachers should have at least a single computer available to download software onto

Table 1 National Educational Technology Standards (NETS)

Basic operations and concepts
- Students demonstrate a sound understanding of the nature and operation of technology systems.
- Students are proficient in the use of technology.

Social, ethical, and human issues
- Students understand the ethical, cultural, and societal issues related to technology.
- Students practice responsible use of technology systems, information, and software.
- Students develop positive attitudes toward technology uses that support lifelong learning, collaboration, personal pursuits, and productivity.

Technology productivity tools
- Students use technology tools to enhance learning, increase productivity, and promote creativity.
- Students use productivity tools to collaborate in constructing technology-enhanced models, preparing publications, and producing other creative works.

Technology communication tools
- Students use telecommunications to collaborate, publish, and interact with peers, experts, and other audiences.
- Students use a variety of media and formats to communicate information and ideas effectively to multiple audiences.

Technology research tools
- Students use technology to locate, evaluate, and collect information from a variety of sources.
- Students use technology tools to process data and report results.
- Students evaluate and select new information resources and technological innovations based on the appropriateness for specific tasks.

Technology problem-solving and decision-making tools
- Students use technology resources for solving problems and making informed decisions.
- Students employ technology in the development strategies for problem solving in the real world.

the PDA. Of course, the computer station needs access to the internet. In addition, it will be helpful to load WinZip, file compression software available at *www.winzip.com,* to decompress certain files. Again, it is my goal to give science teachers several examples of software that is currently available for the academic use of PDAs. Hopefully, Figures 1 and 2 will give a snapshot of such applications.

In addition to the software mentioned above, several other sites are available that offer free PDA shareware. For example, at *www.palm.com/software*, you can access over 10,000 free shareware packages for personal and classroom use. In addition, by downloading Palm Reader to your PDA, you will have access to over 8,000,000 free ebooks, available at *http://etext.lib.virginia.edu/ebooks/* and courtesy of the University of Virginia Library. For student reference, titles such as *Einstein's Principles of Research* are available. However, I must caution that being able to load several books onto a PDA can give students an opportunity to become sidetracked. For example, my PDA contains downloaded

A selection of PDA Software

Periodic Table for Palm OS® v.2.33

Description: Check the elements and compute formulas with this complete table of the periodic elements. You can also compute the weight of formulas after viewing the details of an element.

Price: $10

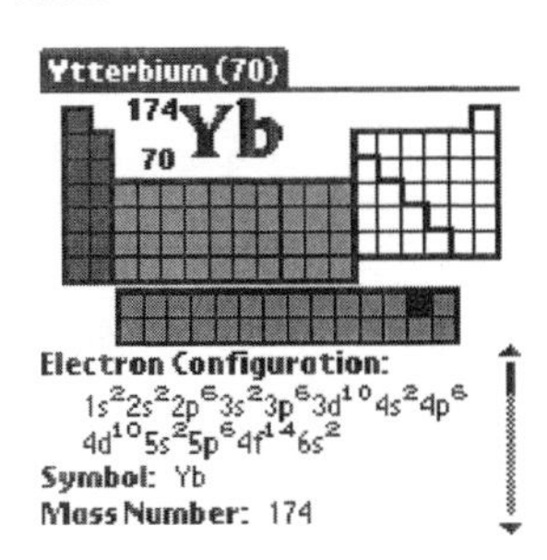

Flash Boom v.1.0

Description: With this application you will be able to estimate the distance to a lightning strike. You can choose between kilometers and miles in the distance display. The cloud icon moves relative to the position of the person icon to indicate the proximity of the lightning strike.

Price: Free (shareware)

CplxCalPro v.2.00 (graphing calculator)

Description: Are you tired of programmable calculators that just look nice? CplxCalPro is the most powerful programmable graphical calculator for the Palm platform. It has more than 175 built-in functions, 30 built-in constants, and eight different display formats. CplxCalPro allows you to work with financial as well as statistical functions, which can be plotted without any programming. It features a fully supported clipboard, autoscaling of graphs, an equation solver worksheet, and much more.

Price: $34.75

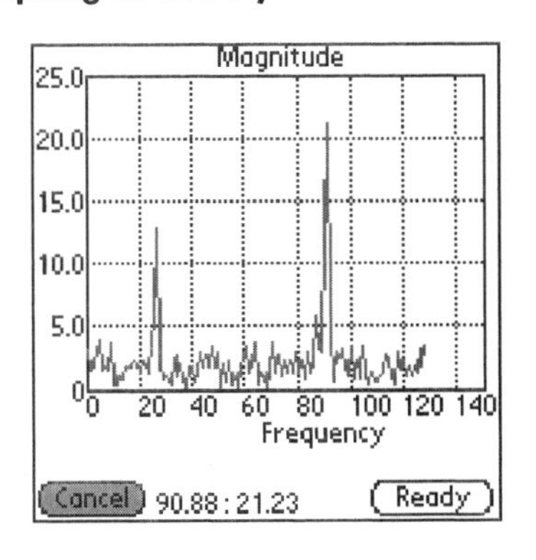

Convert-It! v.1.7 (ART-Personal5.tif)

Description: Converting between different units of measurement is a snap with Convert-It! This software features 41 measurement categories, nearly 500 conversion factors, and an easy-to-use interface.

Price: $10

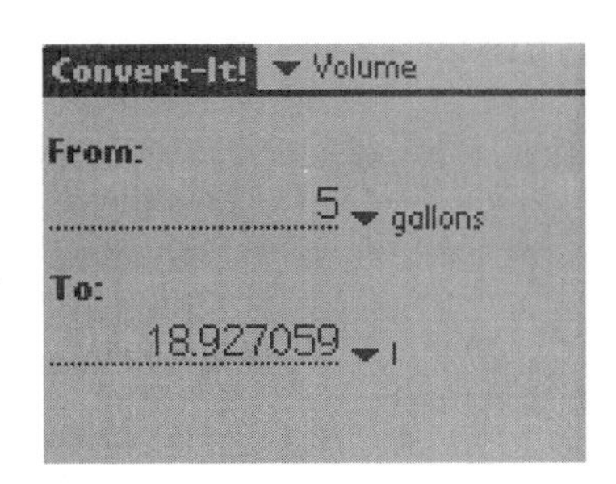

Planetarium v.2.1.1

Description: Planetarium is an application for the beginning stargazer as well as for the professional astronomer. Use it to find or identify objects in a clear nighttime sky. It calculates the position of the Sun, the Moon, the planets, up to 9,000 stars, hundreds of deep sky objects, comets, and asteroids for any time and any geographical position, and draws sky maps of any section of the sky. It shows you twilight times, moon phases, and various astronomical information such as coordinates, magnitude, distance, rise and set times of the planets and all the other objects in the sky.

Price: $31.95

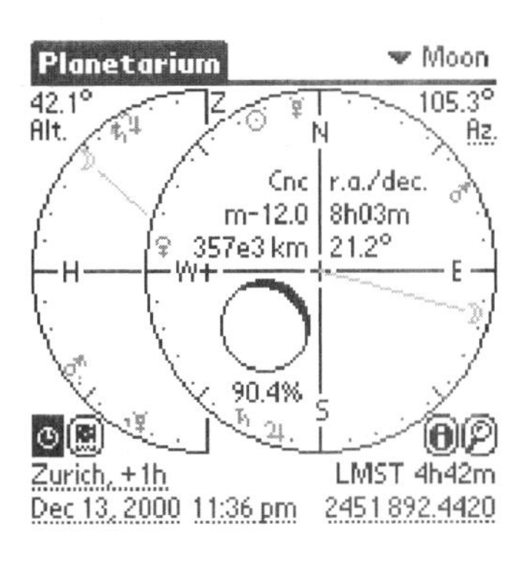

Other PDA Software of Interest

Software Title	Description	Price
Student Log v. 1.0	Keeps track of schoolwork and assignments	$11.50
A-Lex Instant Lookup English Dictionary	General purpose dictionary	$11.95
Steam Properties v. 2.21	Explore properties of steam and water	$20.00
Electrolytes and Nutrition	Notes on metabolism	$12.00
LectureMate v. 1.0	Generate notes and reports	$19.95
SimpleChart v. 1.1	A graphing/plotting application	free
Lesson Plan (PC) v. 3.0	Maintain your lesson plans. Also view, edit, and print them on your desktop PC.	$19.95
Teacher's P.E.T. v. 3	Grading, attendance, and contact info, and more	$19.99
Cycle Timetable	Timetable and school calendar	$12.99
PowerOne Scientific v. 2.0	Full-feature scientific calculator	$29.99

titles by authors like Ellen Glasgow, William Faulkner, and Homer, and can be read using the PDA inconspicuously anywhere (yes, even during faculty meetings!). Therefore, to keep students on task, teachers should monitor PDA usage during class. Otherwise, the costs may outweigh the benefits.

Conclusion

Obviously, for science teachers, the potential for PDA-based software is virtually endless. Hopefully, the ideas presented here will generate additional professional discussion concerning PDA-based classroom software applications for middle school science teachers, as well as to better prepare students for the national and global competition of the 21st century.

References

International Society for Technology in Education (ISTE). 1998. *National educational technology standards for students.* Eugene, OR: ISTE Press.

Palm, Inc. 2001. Palm handheld computers offer simple, quick, smart, and fun ways to teach, learn, and communicate. Press Release. Santa Clara, CA: Author.

Resources

Teachers interested in previewing any of the software mentioned in this article should visit *www.handango.com/palm* to download free demos. This site also provides information on additional titles of interest to middle level teachers.

Projecting Your PDA

EDWIN P. CHRISTMANN

One of the perceived limitations of the PDA by teachers is its small screen. However, some recent innovations have opened a variety of new opportunities for middle school science instruction. Now, with some relatively inexpensive equipment and software, science teachers can give students and opportunity to view PDA applications by projecting whatever is on the PDA's liquid-crystal display (LCD) to a larger viewing screen. Using the same equipment, you can also connect your PDA to a computer to transfer the LCD image to a larger monitor.

What equipment do I need?

To project your PDA screen and make presentations, you will need a PDA (such as a Palm, Sony Clié, Handspring Treo, or Kyocera), a projector (available in a range of sizes and prices, from companies such as InFocus, Proxima, Sharp, and Toshiba) with a standard VGA connector, and a PDA presentation package (see Figure 1 for a list of packages).

PDA presentation software allows you to project anything that you see on your PDA's liquid-crystal display onto a screen for your students to view. So, for example, if you are teaching a lesson on astronomy and have 2sky software loaded on your PDA, by knowing the longitude and latitude of any position, you can project different sky views to your class from any location in the world. It is almost like carrying a portable planetarium in your pocket (see Figure 2). See page 99 for a list of PDA software applications that are suitable for middle school. You can also check the Palm website (*www.palm.com/software*) for other applications.

Conclusion

Obviously, for science teachers, the potential for PDA-based software presentations is virtually endless. Hopefully, the ideas presented here will generate new and innovative ideas concerning the applications of PDAs and PDA software applications for middle school science teachers, as well as to better prepare students for the ever-changing world of the 21st century.

Figure 1 PDA presentation packages

Pitch Solo!
http://www.mobilityelectronics.com/handheld/presentation/pitch-solo.htm
$249.99
Deliver notebook-quality PowerPoint presentations from your Palm Powered device with the new Pitch Solo. PDA is linked to a projector or computer monitor. During slide shows, presentations are controlled remotely using infrared capabilities of your PDA while simultaneously showing your slide notes on your device.

Voyager VGA CF Card for Pocket PC + PC Card
$295.00
www.colorgraphic.net/newsite/products/handheld_overview.asp
Voyager's CF card expands your display capabilities and lets you work with practically all VGA monitors or projectors out there—you can even use a TV monitor thanks to the card's TV-Out feature. The card is also capable of outputting MPEG files, and allows you to set up duplicate and full-screen presentations. CF Voyager VGA is compatible with most presentation software packages.

iPAQ FlyJacket i3800
$249.00
www.lifeview.com
The FlyJacket i3800 can output to a projector, a VGA monitor, or TV directly from an iPAQ. Additionally, the bundled pen-sized remote control/pointer has been specifically designed to perform the task of "Page Up" and "Page Down" for Microsoft PowerPoint file presentations. The opposite end of the device is equipped with a laser diode that serves as a pointer during presentations.

National Education Technology Standards

1. Basic operations and concepts
   - Students are proficient in the use of technology.
2. Social, ethical, and human issues
   - Students practice responsible use of technology systems, information, and software.
   - Students develop positive attitudes toward technology uses that support lifelong learning, collaboration, personal

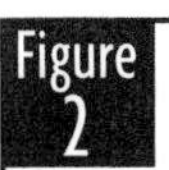
2Sky astronomy software

$25.00
www.handango.com
2sky turns your handheld into a dazzling, dynamic star chart. Take it outside on a clear night. Scroll around the sky with your stylus. Zoom in to see the shapes of galaxies. Animate eclipses with one hand. Transport yourself instantly to any location, date, and time with a few quick taps.
2sky's exclusive "saved view" feature lets you save your anniversary moonrise, or your telescope eyepiece settings for instant recall. 2sky can invert the view in each or both axes and allows you to add a circle to simulate your telescope's field of view.

2sky includes the catalogs of Messier, Caldwell, Flamsteed, Bayer, and the Saguaro astronomical Club, plus 148 objects from Jack Bennett's Southern sky list. It also has lore for all 88 IAU constellations, distances for all but a couple of stars, and 750 searchable star names. If 2sky's 15,560 stars (magnitude 7) and its 500 deep-sky objects aren't enough for you, you can always add optional plug-in databases, such as 2sky Magnitude 11.2 Plug-in, v. 3.0. This software adds 1,231,944 more stars to the 2sky database!

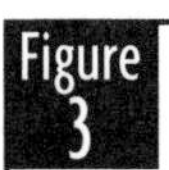
Documents to Go

$49.95
www.dataviz.com/products/documentstogo
If you are a PowerPoint user, a slick software package called Documents to Go can be used to upload files from your computer onto your PDA; or from your PDA onto your computer. Documents to Go can be used to create and edit Word, Excel, and PowerPoint files on your handheld. These documents can then be projected on to a larger screen or monitor to share with your class.

pursuits, and productivity.

3. Technology productivity tools
    - Students use technology tools to enhance learning, increase productivity, and promote creativity.
    - Students use productivity tools to collaborate in constructing technology-enhanced models, prepare publications, and produce other creative works.
4. Technology communications tools
    - Students use telecommunications to collaborate, publish, and interact with peers, experts, and other audiences.
    - Students use a variety of media and formats to communicate information and ideas effectively to multiple audiences.
5. Technology research tools
    - Students use technology to locate, evaluate, and collect information from a variety of sources.
    - Students use technology tools to process data and report results.

Science Research on the Internet

ROBERT A. LUCKING AND EDWIN P. CHRISTMANN

A recent survey by WebTop.com found that 71% of people who use the internet said they were frustrated by web searches and 46% found them nerve-racking. Young students in particular often get frustrated trying to find useful information for school purposes and end up jumping from one point of interest to another, keeping little focus on the task at hand. To help students achieve success in their searches, teachers should direct them toward proven search engines and other online resources. The following are some of my recommendations.

Search with success

A great place to send kids to conduct their science searches is the Why Files at *http://whyfiles.org* (see Figure 1). Search hundreds of articles about "the science behind the news" of current events from the University of Wisconsin-Madison Graduate School. The Why Files attempts to portray science as a critical human endeavor conducted by ordinary people, using news and current events as springboards to explore science, health, the environment, and technology. An authority in the field in question checks all of the articles, and fact-checkers are listed.

If your students are more sophisticated in their search experiences, you may want to move them on to what is probably the most comprehensive search engine available to those interested in the sciences: Scirus, located at *www.scirus.com* (see Figure 2). Driven by the latest in search engine technology, Scirus claims to cover over 105 million science-related pages, consisting of 90 million websites and 17 million records from many special-emphasis sources that may be of interest to students or teachers such as ScienceDirect, MEDLINE on BioMedNet, Beilstein on ChemWeb, Neuroscion, BioMed Central, and NASA.

All of these offer potentially useful information, including peer-reviewed articles by scientists and scholars at the top of their fields, and it is possible to choose a search based on websites or journal articles or both. Under their advanced search option, it is possible to choose from various search engines within individual disciplines, or to work within subject domains (by using any number of check boxes for the 17 subjects available). In the scientific world you will often find papers in PDF format, and Scirus contains additional capabilities to search PDF documents and therefore to locate information that's often invisible to other search engines.

A second sophisticated search site, CiteSeer, at *http://citeseer.ist.psu.edu* (see Figure 3), is a digital, high-end scientific library designed to improve the dissemination and feedback of scientific literature. CiteSeer, computes citation statistics and related documents for all articles cited in its database, not just the indexed articles. This is an ideal tool for serious researchers.

Figure 1 Why Files (whyfiles.org)

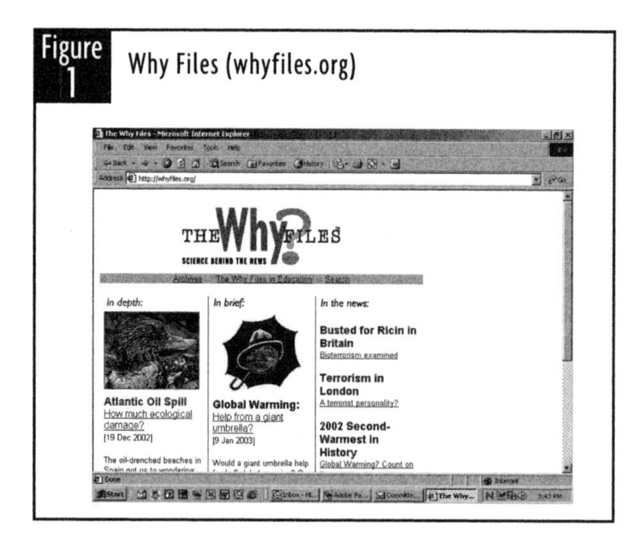

Figure 2 Scirus (www.scirus.com)

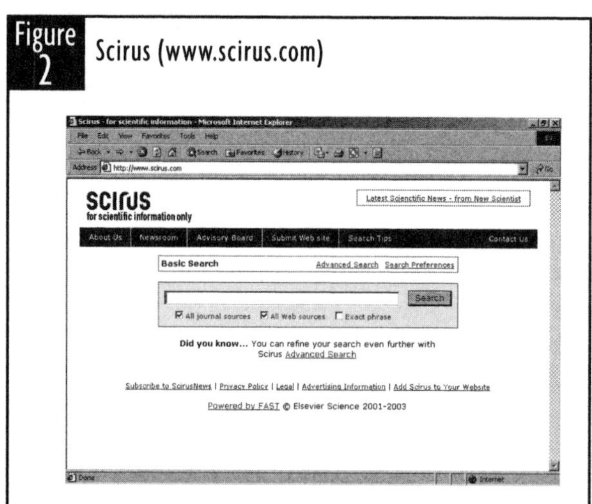

Figure 3 CiteSeer (citeseer.com)

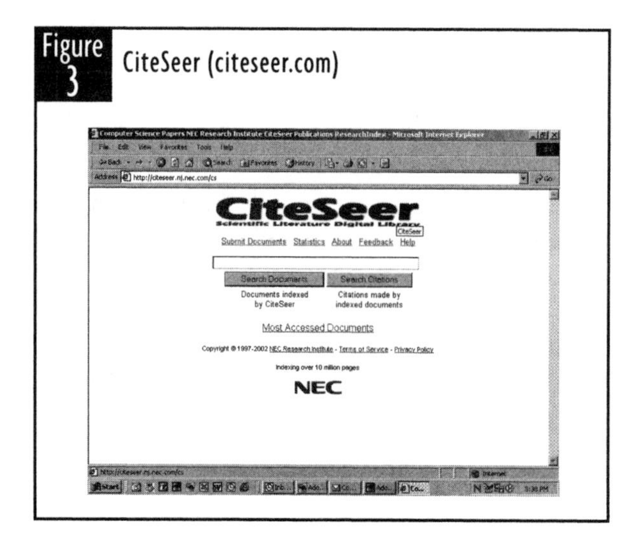

An interesting problem your students may face is that they may never have a chance to see sites or information uncovered by the search engine. While space doesn't allow for a full exploration of the topic of censorship, mere access to useful information does become an issue in classrooms. In many schools, censorware may limit the type or amount of information students may visit on the internet. Net Nanny, Cybersitter, Cyber Snoop, Guardian, GuardiaNet, Surf Watch, Cyber Patrol, The Internet Filter, Net Shepherd, and Smart Filter are just a few of the various internet-filtering companies, many of which sell their services to schools.

Another solution is kid-friendly search engines. These include Ask Jeeves for Kids, at *www.ajkids.com* (see Figure 4), Yahooligans, at *http://yahooligans.yahoo.com* (see Figure 5), and KidsClick, at *http://sunsite.berkeley.edu/KidsClick!*. Likewise, if you want to feel comfortable recommending searches accessible from students' homes that are reputed to be balanced in their approaches to sensitive topics, Zeeks (*www.zeeks.com*), Headbone Zone (*www.headbone.com*), and Lycoszone (*www.lycoszone.com*) are examples of portal sites offering filtered search engines (as well as chat rooms and safe e-mail).

For your own search interests, you might want to check out one of several commercial search tools that offer many advantages to users in terms of how they carry out their searches and manage the results of the search. In the interest of space, we will review just one, Copernic, at *www.copernic.com* (see Figure 6). Three different versions of this software are available, ranging in features represented by a free download to a $79 version known as Copernic Agent Professional. Although this latter version is pricey, we will consider its many capabilities because teachers might choose to use it with students as well as for their own personal or educational purposes.

Copernic acts much like a metasearch engine because it boasts of being capable of searching 1,000 search engines in 120 different categories. Its results are broad, leaving you in control of the number of returns, and significantly removing duplicate hits. Its strength rests in the search results management capabilities that allow you to view a good deal of information about each returned hit, sort the results, and, most importantly, keep a history of your searches using a system of folders. When your returns are displayed in a window on the screen, you are able to see information about each site identified in your search, and you can then search the results of your search to further refine your focus! We find that one of the greatest advantages of software of this sort is its ability to save the results of a search. Typically, we jump back and forth between various articles found in a search, uncertain of exactly what we want until we have read several pieces. Copernic keeps the results clearly accessible, and you can save the results in a system of files. Additionally, you can export your search results to a Word file, meaning that you can carry a disk that contains any search results you choose to another computer at another location.

If you have been around the internet for a long time, you may remember when a subset of computer users burrowed into an arcane collection of postings by enthusiasts of every conceivable interest. Usenet has now grown up, and it comprises hundreds of online "bulletin boards." More than 13,000 groups have been formed to discuss topics that span every human endeavor imaginable. People from around the world access messages posted to individual groups. In many cases, it is an appropriate forum for asking questions, particularly if you're looking for broad or technical discussion on a given topic.

Google Groups (Figure 7) contains the entire archive of Usenet discussion groups dating back to 1981, allowing you to conduct searches providing relevant results from a database containing more than 700 million posts (*http://groups.google.com/advanced_group_search*). Under the science option, over 175 different groups are listed. And Google has another card up its sleeve, especially if you're a Microsoft Internet Explorer for Windows user (see sidebar). Here you can download

Figure 4 Ask Jeeves for Kids (www.ajkids.com)

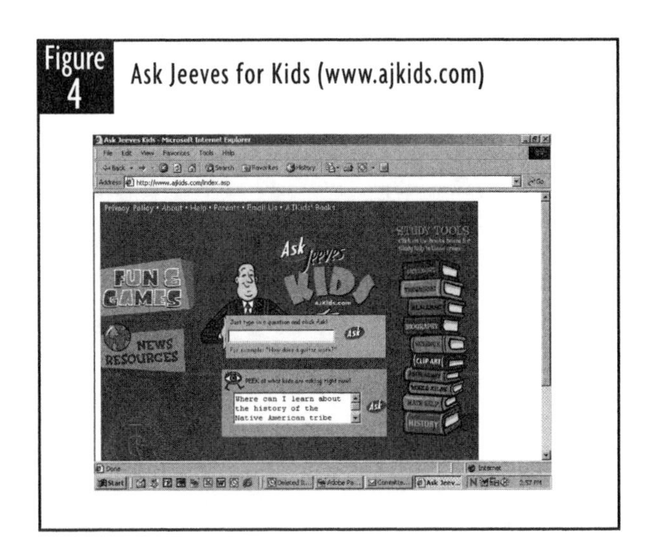

Figure 5 Yahooligans (www.yahooligans.com)

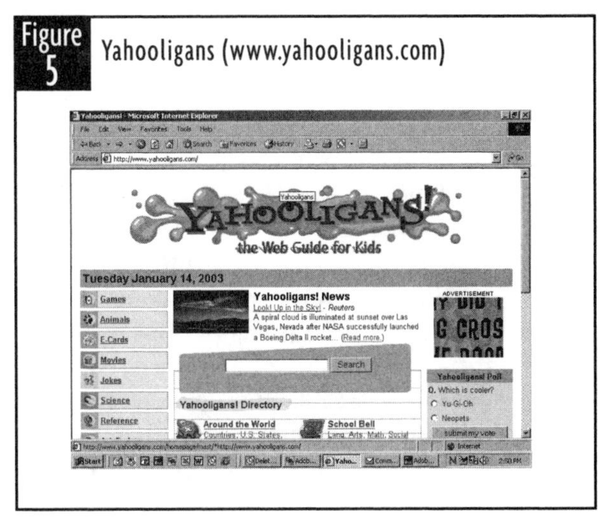

Figure 6 Copernic Agent Professional

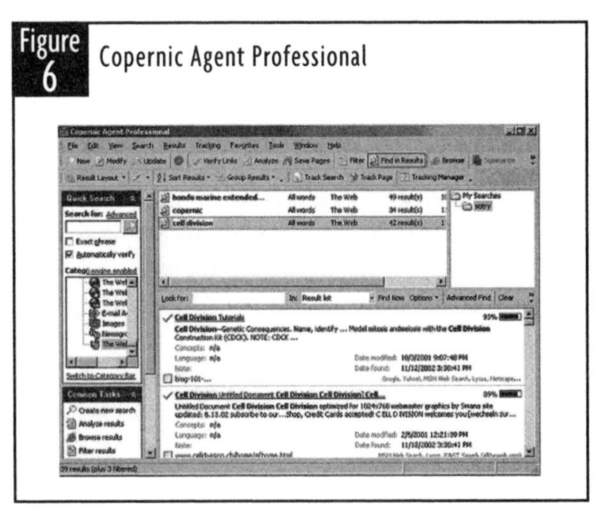

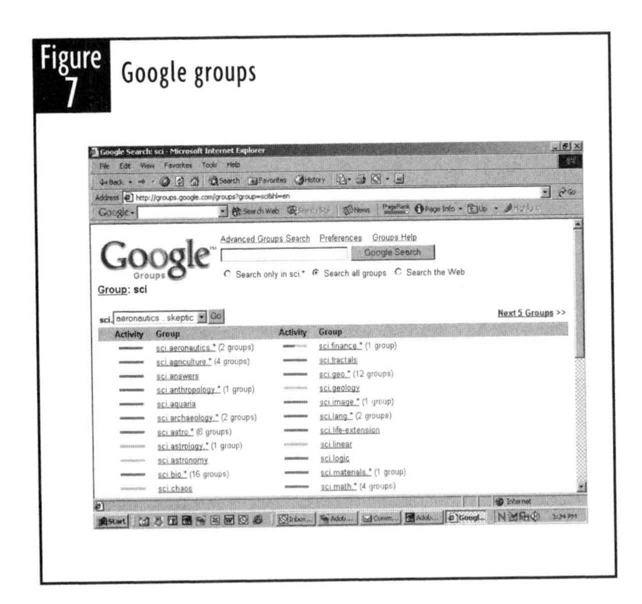

Figure 7 Google groups

a toolbar that integrates directly with the search engine you may be using.

Once downloaded, the program creates a small Google toolbar on IE (version 5.0 or later). Typing a word or phrase into the toolbar and clicking the "search" button brings up a list of relevant websites, just like the one produced with the search engine itself. However, if you click the toolbar's "search site" button instead, your search will be limited to the site you are on—often with better results than using that site's own search capabilities. The toolbar also gives access to other Google features, including advanced searches that include many other options.

National Standards

Internet-based research generates activities that are aligned with Content Standard A: Science as Inquiry, "Students in grades 5–8 should be provided opportunities to engage in full and in partial inquiries;" and Content Standard E: Science and Technology, "...students' work with scientific investigation can be complemented by activities that are meant to meet a human need, solve a human problem, or develop a product...." for students in grades 5–8 (NRC 1996).

The integration of Internet research into science instruction conforms to the tenets of the National Education Technology Standards for Students (ISTE 1998) as follows:

1. Basic operations and concepts
   - Students demonstrate a sound understanding of the nature and operation of technology systems
   - Students are proficient in the use of technology
2. Social, ethical, and human issues
   - Students understand the ethical, cultural, and societal issues related to technology
   - Students practice responsible use of technology systems, information, and software
   - Students develop positive attitudes toward technology uses that support lifelong learning, collaboration, personal pursuits, and productivity
3. Technology productivity tools
   - Students use technology tools to enhance learning, increase productivity, and promote creativity
   - Students use productivity tools to collaborate in constructing technology-enhanced models, prepare publications, and produce other creative works
4. Technology communications tools
   - Students use telecommunications to collaborate, publish, and interact with peers, experts, and other audiences
   - Students use a variety of media and formats to communicate information and ideas to multiple audiences
5. Technology research tools
   - Students use technology to locate, evaluate, and collect information from a variety of sources
   - Students use technology tools to process data and report results
   - Students evaluate and select new information resources and technological innovations based on the appropriateness for specific tasks

6. Technology problem-solving and decision-making tools
    - Students use technology resources for solving problems and making informed decisions
    - Students employ technology in the development of strategies for solving problems in the real world

Conclusion

Obviously, the potential for internet-based research is endless. Additional professional discussion concerning Internet research applications for middle school science teachers is encouraged, and the focus of that conversation must be the preparation of students for the national and global dynamics of the 21st century.

Internet resources

Why Files:
http://whyfiles.org
Scirus:
www.scirus.com
CiteSeer:
http://citeseer.ist.psu.edu
Jeeves for Kids:
www.ajkids.com
Yahooligans:
www.yahooligans.com
Kids Click:
http://sunsite.berkeley.edu/KidsClick!
Google Groups:
www.groups.google.com

References

International Society for Technology in Education (ISTE). 1998. *National educational technology standards for students.* Eugene, OR: ISTE Press.

National Research Council (NRC). 1996. *National science education standards.* Washington, DC: National Academy Press.

Richardson, C. R, P. J. Resnick, D. L. Hansen, H. A. Derry, and V. J. Rideout. 2002. Does pornography-blocking software block access to health information on the internet? *Journal of the American Medical Association* 288: 2887–2894.

Google Toolbar (http://toolbar.google.com)
The Google Toolbar increases your ability to find information from anywhere on the web and takes only seconds to install. When the Google Toolbar is installed, it automatically appears along with the Internet Explorer toolbar. This means you can quickly and easily use Google to search from any website location, without returning to the Google home page to begin another search.

The Google Toolbar is available free of charge and includes these great features:

- **Google Search:** Access Google's search technology from any web page.
- **Search Site:** Search only the pages of the site you're visiting.
- **PageRank:** See Google's ranking of the current page.
- **Page Info:** Access more information about a page including similar pages, pages that link back to that page, as well as a cached snapshot.
- **Highlight:** Highlight your search terms as they appear on the page; each word in its own color.
- **Word Find:** Find your search terms wherever they appear on the page.

Technologies for Special Needs Students

EDWIN P. CHRISTMANN AND ROXANNE R. CHRISTMANN

In 1997, the Individuals with Disabilities Education Act Amendments (PL 105-17) ensured that students with disabilities have access to general education. This act adheres to the notion that parents, students, and teachers will work together to design an individualized education program (IEP) for special needs students. As a result, the inclusion of special needs students into regular classes has been mandated. Therefore, it is essential that science teachers have an understanding of the variety of technologies that are available for special education students. This is especially important, considering that classroom science teachers are legally required to participate in the development of a student's IEP. An IEP is a document that specifies guidelines for modifications to a student's classroom instruction. These modifications are based on the special needs of a student and should be designed to increase the probability of classroom success for the special needs student.

You might wonder why today's science teachers need to keep current with the latest technologies available for special education students. When you consider that more than 50 million Americans are identified as having a disability, classroom technology applications for special education students is crucial (Sharp 2002). Especially for those who would generally not be able to participate in scientific activities without the use of the latest technologies. Moreover, with the federal government's mandate of the Individuals with Disabilities Education Act Amendments of 1997, it is imperative that science teachers comply with the state and federal special education guidelines.

Computer-assisted instruction

Research supports the use of computer-assisted instruction (CAI) for special needs students as a supplement to traditional instruction (Christmann, Badgett, and Lucking 1997). One of the obvious benefits is that a computer allows special needs students to work at an individual pace. Several excellent science-based CAI software packages are available for special needs students. An excellent example is Edmark's Virtual Lab Series, which gives special education students the opportunity to explore light and electricity *(http://store.sunburst.com/ProductInfo.aspx?itemid=176409).* Through computer simulation software, students can participate in lab activities that might otherwise be difficult, if not impossible. In addition, students who are unable to perform tasks that require the use of fine motor skills can use software that operates with single-switch technology. Single-switch technology allows students to trigger mouse "clicks" without applying pressure to mouse buttons. Another software designer, DK Multimedia, has several science programs, such as Earth Quest, Nature 2.0, and Dinosaur Hunter, that incorporate multimedia software applications into instruction for special needs students *(www.educate-me.net/educate/category.cfm?Category=126).*

Assistive technologies

Another form of technology that is available for students with disabilities is *assistive technology* (AT). Basically, anything that makes a task easier to perform (including handheld text readers, sonar vision glasses for the blind, and keyboard aids) is considered assistive technology. Assistive technology also includes services for evaluation, design, customization, adaptation, maintenance, repair, therapy, training, or technical assistance (Sharp 2002).

On a daily basis, all people use technology to function more fully in their lives. However, for people who have disabilities, it is sometimes impossible to function in a world designed for people without disabilities. Ironically, Stephen Hawking, a world famous scientist, has benefited from some of the same assistive technologies that are available for students today (see Figure 1 for examples). Because disabilities differ among students, each student must be fitted with assistive technologies that are commensurate with their individual needs. Therefore, we have included a list of internet resources for your future reference.

Conclusion

It is also important to note that in order to stay in compliance with the Individuals with Disabilities Education Act Amendments of 1997, it is imperative for science teachers to work directly with special education teachers to incorporate assistive technologies into students' science instruction based on individual needs. Currently, there are many technology applications for special education instruction. These applications allow special education students to feel comfortable and experience more success in the classroom, which are two key elements of any successful learning environment.

Internet resources

Assistive Technology Solutions:
www.abilityhub.com

Keyboard for students with special needs:
www.maltron.com/#singlehanded

Special keyboards:
www.fentek-ind.com/bigkey.htm

Special needs software:
http://college.hmco.com/education/resources/res_prof/students/spec_ed/tech_resources/index.html

Voice controlled devices:
www.donjohnston.com

Voice recognition system:
www.intellitools.com

Adaptive Technology Resource Center:
www.utoronto.ca/atrc/reference/tech/techgloss.html

Resources for students with disabilities:
www.disabilityresources.org

References

Christmann, E. P., J. L. Badgett, and R. Lucking. 1997. The effectiveness of microcomputer-based computer-assisted instruction on differing subject areas: A statistical deduction. *Journal of Educational Computing Research* 16(3): 281–296.

International Society for Technology in Education (ISTE). 1998. *National educational technology standards for students.* Eugene, OR: ISTE Press.

National Research Council (NRC). 1996. *Na-*

tional science education standards Washington, DC: National Academy Press.

Sharp, V. 2002. *Computer education for teachers.* New York, NY: McGraw Hill Publishing.

Standards

Special education applications can be aligned with any content standard for students in grades 5–8 (NRC 1996). The integration of internet research into science instruction conforms to the tenets of the National Education Technology Standards for Students (ISTE 1998) as follows:

1. Basic operations and concepts
   - Students demonstrate a sound understanding of the nature and operation of technology systems
   - Students are proficient in the use of technology
2. Social, ethical, and human issues
   - Students understand the ethical, cultural, and societal issues related to technology
   - Students practice responsible use of technology systems, information, and software
   - Students develop positive attitudes toward technology uses that support lifelong learning, collaboration, personal pursuits, and productivity
3. Technology productivity tools
   - Students use technology tools to enhance learning, increase productivity, and promote creativity
   - Students use productivity tools to collaborate in constructing technology-enhanced models, prepare publications, and produce other creative works
4. Technology communications tools
   - Students use telecommunications to collaborate, publish, and interact with peers, experts, and other audiences
   - Students use a variety of media and formats to communicate information and ideas to multiple audiences
5. Technology research tools
   - Students use technology to locate, evaluate, and collect information from a variety of sources
   - Students use technology tools to process data and report results
   - Students evaluate and select new information resources and technological innovations based on the appropriateness for specific tasks
6. Technology problem-solving and decision-making tools
   - Students use technology resources for solving problems and making informed decisions
   - Students employ technology in the development of strategies for solving problems in the real world

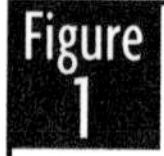

Figure 1 Assistive Technology Glossary

Onscreen keyboards

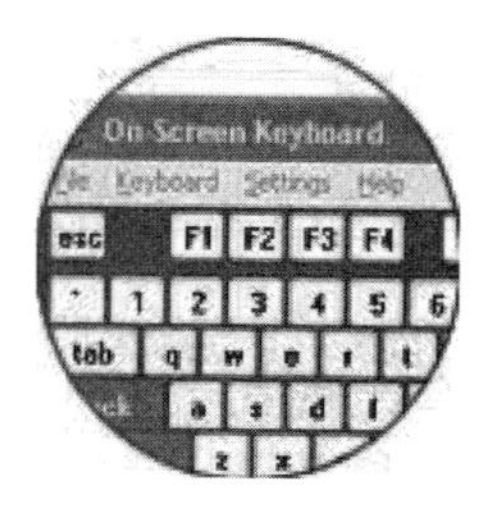

A great number of people are unable, for various reasons, to use a standard keyboard. These reasons can range from limited movement or motor control to low strength in the hands and fingers. For many, an alternative keyboard can solve these problems, but for others the best solution is to use a virtual or onscreen keyboard. An onscreen keyboard generally appears on the same display used for programs and will remain permanently visible. The keyboard can then be accessed using the pointer device. In the simplest sense this means a standard mouse, but through the use of alternative pointer devices or switches a large number of disabilities can be addressed.

Voice recognition systems

Voice recognition allows a user to use his/her voice as an input device. Voice recognition may be used to dictate text into the computer or to give commands to the computer (such as opening application programs, pulling down menus, or saving work).

Alternative keyboards

Alternative keyboard layouts and other enhancements allow people who experience difficulty with conventional keyboard designs to use computers. The products available range from keyguards that prevent accidental key activation, to alternative keyboards with differing layouts, and sizes, for people who have specific needs, to alternative input systems that require other means/methods of getting information into a computer.

Alternative mouse devices

Alternative pointing devices are used to replace the mouse. The keyboard keypad can function as a mouse using mousekeys. Mousekeys allow the user to manipulate the cursor on screen using keys on the keyboard. Many of the alternative keyboards have mousekeys built in, so the keys on the keyboard can toggle between text input or mouse input. Trackballs are upside down mice, with the ball on top and several buttons. Many trackballs offer the left and right mouse buttons plus one or two more that can be programmed to be a double click or drag lock. Many local computer vendors stock trackballs and the programmable ones are also available from assistive technology vendors. These allow the user to manipulate up to five switches to control the mouse—the more switches the user can control, the faster the mouse can be manipulated. Mouse input can also be given by high-tech pointing devices, which transmit the location of a transmitter or reflective dot on the user's head to the computer system. Separate switches, or just dwelling on a location, are used for mouse clicks and drags. These are frequently used with on-screen keyboards for text input by people with limited movement due to quadriplegia or muscular dystrophy.

Speech synthesizers

An external speech synthesizer is a hardware device used for speech output. Typically, they are used with screen readers or optical character recognition/scanning software programs for people who are blind or visually disabled. External speech synthesizers were used exclusively before the advent of sound cards in computers. Now, with multi-channel sound cards people who use screen readers or other speech output software can have both the "voice" of the computer and the system sounds audible at the same time. Some people who require the system sounds, or who prefer to leave their sound cards to perform other functions, may want to use an external speech synthesizer instead of the internal sound card. For example, if they want to listen to a CD, watch a DVD, or do some audio/video conferencing, they might want to leave the sound card channels free to do so. This may also be a critical piece of equipment for people who are also composing music or using audio editing programs.

Screen magnifiers

Screen magnification software is used by people with visual disabilities to access information on a computer screen. The software enlarges the information on the screen by pre-determined incremental factor. Magnification programs run simultaneously and seamlessly with the computer's operating system and applications. Most screen magnification software has the flexibility to magnify the full screen or parts of the screen or to provide a magnifying glass view of the area around the cursor or pointer. These programs often allow for inverted colors, enhanced pointer viewing, and tracking options.

Switches

Switches are a common solution for users with mobility disabilities, such as muscular dystrophy, multiple sclerosis, cerbral palsy, spinal cord injuries, and head injuries, who need to use computers or other electronic devices, but have difficulty with the physical interface. To allow easier manipulation than a standard keyboard or joystick, a specially designed switch may be composed of a single button, merely a few buttons, a sensory plate, or another of the whole host of adaptive switches available; it may also be touch-free, relying instead on motion sensors, brain activation, or a sip-and-puff mechanism. Aside from simplifying user interfaces, switches can also be used as developmental aids, teaching children (or adults) how to interact with their external environment.

Screen reading and talking browsers

A screen reader is the commonly used name for voice output technology. Hardware and software produce synthesized voice output for text displayed on the computer screen, as well as for keystrokes entered on the keyboard. Voice-based browsers use the same technology as screen reading software, but are designed specifically for internet use.

Software That Makes the Grade

EDWIN P. CHRISTMANN

As you make plans for the upcoming school year, consider reviewing one of the many gradebook software packages that are available for organizing and computing your classroom grades. Many different packages are available (see Figure 1), which range in price from free to over $100. The initial set-up time for gradebook software is about the same as setting up a traditional course gradebook. You have to enter students' names and grades on assignments into the corresponding rows and columns.

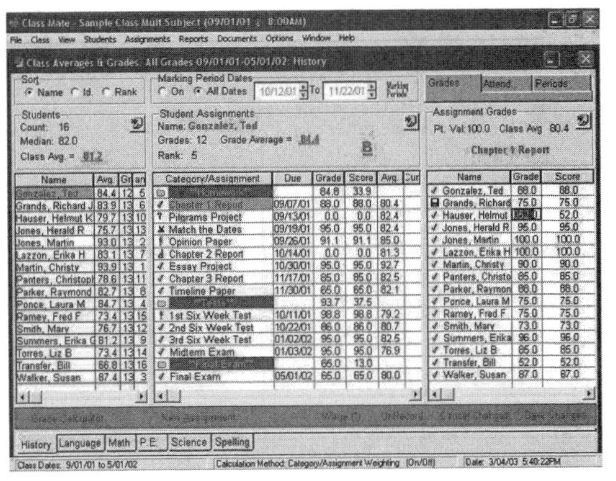

Class Mate main screen *(www.classmategrading.com)*

The advantages of a computer gradebook

Computer gradebooks enable you to track and organize tests, homework assignments, lab work, and other graded assignments. In addition, you can generate seating charts and organize classroom records (such as absences and discipline incidents). You can also generate student progress reports at any time during the marking period, so the reports can be distributed to parents and other school personnel prior to conferences.

Perhaps the greatest benefit to teachers is time saved. The time-consuming task of calculating grades is performed by the software, allowing the teacher to focus on more important matters. And the grading system can be customized to your needs. Student grades can be weighted, curved, or customized to your particular grading scale; and information can be entered as percentages, points, or even letter grades. Extra credit and makeup work can also be accounted for by the software. Computer gradebooks also minimize computational errors by reducing the number of calculations necessary to compute grades. However, keep in mind that a data entry error will result in an incorrect grade. Users need to be careful when entering their data—garbage in, garbage out.

Today's gradebook software packages permit you to import data from word processing and spreadsheet software. For example, if your school provides you with a class roster on a spreadsheet, you can load your class rosters into your computer gradebook instead of typing the information in yourself.

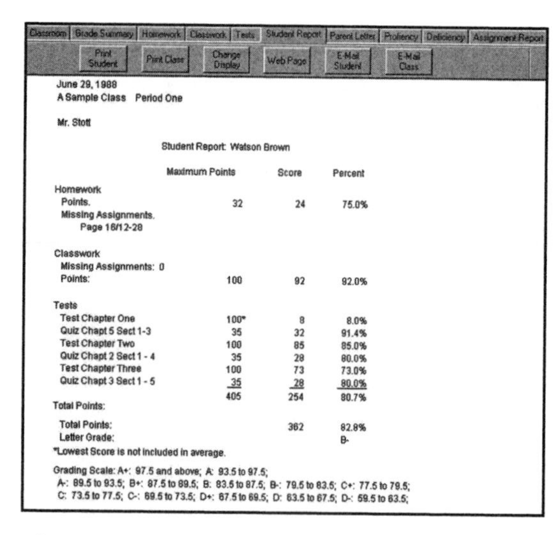
Classroom | Grade Summary | Homework | Classwork | Tests | Student Report | Parent Letter | Proficiency | Deficiency | Assignment Report

Print Student | Print Class | Change Display | Web Page | E-Mail Student | E-Mail Class

June 29, 1988
A Sample Class Period One

Mr. Stott

Student Report: Watson Brown

	Maximum Points	Score	Percent
Homework			
Points.	32	24	75.0%
Missing Assignments.			
Page 16/12-28			
Classwork			
Missing Assignments: 0			
Points:	100	92	92.0%
Tests			
Test Chapter One	100*	8	8.0%
Quiz Chapt 5 Sect 1-3	35	32	91.4%
Test Chapter Two	100	85	85.0%
Quiz Chapt 2 Sect 1 - 4	35	28	80.0%
Test Chapter Three	100	73	73.0%
Quiz Chapt 3 Sect 1 - 5	35	28	80.0%
	405	254	80.7%
Total Points:			
Total Points:		362	82.8%
Letter Grade:			B-

*Lowest Score is not included in average.

Grading Scale: A+: 97.5 and above; A: 93.5 to 97.5;
A-: 89.5 to 93.5; B+: 87.5 to 89.5; B: 83.5 to 87.5; B-: 79.5 to 83.5; C+: 77.5 to 79.5;
C: 73.5 to 77.5; C-: 69.5 to 73.5; D+: 67.5 to 69.5; D: 63.5 to 67.5; D-: 59.5 to 63.5;

Class Action Gradebook summary page
(*www.classactiongradebook.com*)

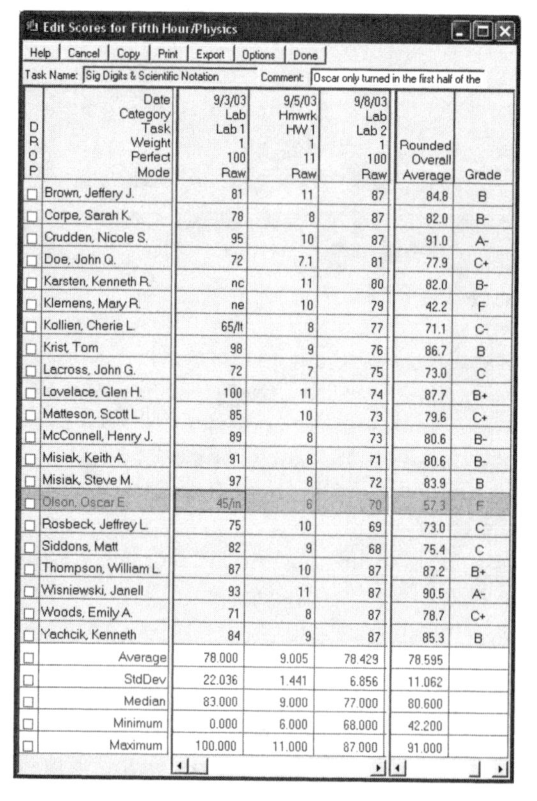
Edit Scores for Fifth Hour/Physics

Help | Cancel | Copy | Print | Export | Options | Done

Task Name: Sig Digits & Scientific Notation Comment: Oscar only turned in the first half of the

DROP	Date / Category / Task / Weight / Perfect / Mode	9/3/03 Lab Lab 1 1 100 Raw	9/5/03 Hmwrk HW 1 1 11 Raw	9/8/03 Lab Lab 2 1 100 Raw	Rounded Overall Average	Grade
☐	Brown, Jeffery J.	81	11	87	84.8	B
☐	Corpe, Sarah K.	78	8	87	82.0	B-
☐	Crudden, Nicole S.	95	10	87	91.0	A-
☐	Doe, John Q.	72	7.1	81	77.9	C+
☐	Karsten, Kenneth R.	nc	11	80	82.0	B-
☐	Klemens, Mary R.	ne	10	79	42.2	F
☐	Kollien, Cherie L.	65/lt	8	77	71.1	C-
☐	Krist, Tom	98	9	76	86.7	B
☐	Lacross, John G.	72	7	75	73.0	C
☐	Lovelace, Glen H.	100	11	74	87.7	B+
☐	Matteson, Scott L.	85	10	73	79.6	C+
☐	McConnell, Henry J.	89	8	73	80.6	B-
☐	Misiak, Keith A.	91	8	71	80.6	B-
☐	Misiak, Steve M.	97	8	72	83.9	B
☐	Olson, Oscar E.	45/in	6	70	57.3	F
☐	Rosbeck, Jeffrey L.	75	10	69	73.0	C
☐	Siddons, Matt	82	9	68	75.4	C
☐	Thompson, William L.	87	10	87	87.2	B+
☐	Wisniewski, Janell	93	11	87	90.5	A-
☐	Woods, Emily A.	71	8	87	78.7	C+
☐	Yachcik, Kenneth	84	9	87	85.3	B
☐	Average	78.000	9.005	78.429	78.595	
☐	StdDev	22.036	1.441	6.856	11.062	
☐	Median	83.000	9.000	77.000	80.600	
☐	Minimum	0.000	6.000	68.000	42.200	
☐	Maximum	100.000	11.000	87.000	91.000	

School Maestro III score edit screen
(*www.rredware.com*)

What package is best for you?

The best ways to choose a computer software package are to order a trial version or download one from the internet. For example, for a $4 shipping fee, Russ and Ryan EdWare will send you a trial version of their gradebook software, School Maestro III. You also could download a trial version of ThinkWave Educator at *www.thinkwave.com/productseducatortrial.html*. To determine which package works best for you, obtain a trial version and then attempt to use all of the features that you would like to use to manage your science classroom. You may have to compare a few packages before finding one that fits your needs. Preview as many packages as possible before you buy. You should also ask around your school to see what other packages teachers are using. See what they like and ask for a hands-on tutorial if possible. Figure 1 lists a number of packages and provides internet links to each publisher's homepage.

Closing the book

Because many school districts have not mandated the use of gradebook software, science teachers can select their own software package for calculating grades and managing their classrooms. Undoubtedly, gradebook software can help you to maximize your instructional time by minimizing the time you spend calculating grades and writing student progress reports. For the upcoming school year, consider adopting a tool that can help you to make the most out of each school day.

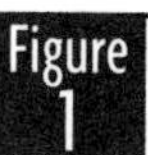

Gradebook Software Packages

1st Class Software, 1st Class GradeBook—*www.1stclasssoftware.com/default.asp*
*CalEd Software, Class Action Gradebook—*www.classactiongradebook.com*
CampusWare, GradeSpeed—*www.gradespeed.net/Campusware/gs/index.htm*
*Chariot Software, MicroGrade—*www.chariot.com/micrograde/index.asp*
*Class Mate Software, Class Mate Grading Software—*www.classmategrading.com*
*ClassBuilder Software, ClassBuilder—*www.classbuilder.com*
Common Goal Systems, TeacherEase—*www.teacherease.com*
E-Z Grader, E-Z Grader—*www.ezgrader.com*
EdSoft Software, Instructional Management System—*www.ed-soft.com*
*ESembler, eSembler for Education—*www.esembler.com*
*Excelsior Software, Pinnacle—*www.excelsiorsoftware.com*
*Jackson Software, GradeQuick—*www.jacksonsoftware.com*
*Misty City Software, Grade Machine—*www.mistycity.com*
*Russ and Ryan EdWare, GradeBook for Windows—*www.rredware.com*

* Indicates that a free trial version is available.

Computer-Assisted Instruction

EDWIN P. CHRISTMANN

Throughout the nation, schools are incorporating computer-assisted instruction (CAI) into their science curricula in efforts to enhance student achievement. CAI has been divided into several subcategories (Atkinson 1969, Watson 1972). For example, computers can supplement traditional drill-and-practice with relevant practice exercises. The tutorial mode of the computer presents the student with an introduction of concepts that is followed by appropriate questioning strategies. Simulations allow students to assume roles that motivate them toward the accomplishment of realistic goals. As an instrument to problem solving, the computer assists students in making decisions, following logical steps, and finding answers to problems through the use of provided information. Through its provision of educational games, the computer allows the student to learn through entertaining and recreational activities. Moreover, its problem-solving format is conducive to the creation of higher-level skills, tailored to the specific needs of all students, across several disciplines (Mayes 1992).

Research substantiates that CAI is an effective form of supplemental instruction at both the elementary and secondary school levels (Christmann 1997). Not only are CAI instructional methods successful in enhancing achievement in various subject areas, but also they have been shown to be comparatively more effective in raising the academic achievement of younger students, lower-ability students, and disadvantaged students. However, some research suggests that boys have greater confidence and higher levels of interest in using computers. Moreover, academic achievement gains were found to be higher among boys when CAI drill-and-practice was used as an instructional method. Research also indicates that academic achievement gains can especially be realized when traditional instruction is supplemented with CAI for 15 to 20 minutes per day, four days per week (Becker 1986).

The literature is legion in its number of studies confirming the effectiveness of CAI in the advancement of academic achievement. Currently, educational researchers are focusing their efforts on the effects of such instruction in differing educational settings, perhaps in agreement with the findings of Mauriel (1989), which disclose that the essence of student differences lies in differing social, economic, demographic, and educational backgrounds (Christmann, Lucking, and Badgett 1997). Because each subject and grade level is driven by unique curricular and instructional characteristics, the effectiveness of computer-assisted instruction varies on the basis of student differences.

CAI and achievement

The mean effect size calculated for science students who supplemented traditional science instruction with CAI is 0.266, indicating that the average student exposed to CAI showed academic achievement that was greater than that of 60.4 percent of those students who were exposed to traditional instruction. Moreover, the typical student moved from the 50th percentile to the 60.4th percentile when exposed to CAI (Christmann and Badgett 1999).

When CAI was compared with traditional instruction (Christmann, Badgett, and Lucking 1997), the largest mean effect size was found in science (0.639), which indicates that the average science student exposed to CAI attained academic achievement greater than that of 73.9 percent of those science students exposed to traditional instruction. This supports other research that shows CAI as having its greatest effect in science (Roblyer 1985).

A plausible explanation for this is that the structure of science is commensurate with the discrete and objective steps that are inherent in CAI instruction. Moreover, microcomputer simulations enable students to learn science through actual experiences rather than the vicarious method of reading about experiments. Here, students can conduct experiments that are normally considered dangerous, impossible, or impractical. For example, simulations can allow for an analysis of the genetic characteristics of many generations within a single laboratory session. Other examples, such as experiments with diffusion, osmosis, mitotic division, and population problems can be simulated by microcomputers in a very short period of time at a nominal expense. Such simulations free students from time-consuming procedures so they can concentrate on the comprehension and mastery of the material.

Clearly, CAI software offers science teachers a kaleidoscope of possible instructional applications to keep students competitive in a contemporary world. Consequently, in an era when CAI software has moved from mainframe-based computers to classroom-based microcomputers, teachers need to gain a better awareness of how microcomputer-based CAI software can be used in their classrooms. Moreover, as the number of microcomputers used for instructional purposes increases, it seems increasingly important to use them in the most effective and productive manner. In response to this dilemma, *Science Scope* plans to dedicate several Tech Trek columns to specific CAI software packages and applications that can be integrated into middle school science instruction. Hopefully, this will help clarify the availability of CAI software, as well as, the efficient and effective use of the most updated CAI technology available for teachers today.

References

Atkinson, R. 1969. Computerized instruction and the learning process. New York: Academic Press.

Becker, J. 1986. Instructional uses of school computers: Reports from the 1985 national survey. Baltimore: John Hopkins University, Center for the Social Organization of Schools.

Christmann, E. P. 1997. The effectiveness of computer-assisted instruction: What research tells us. West Lafayette, IN: Kappa Delta Pi.

Christmann, E. P., and J. L. Badgett. 1999. A comparative analysis of the effect of computer-assisted instruction on student achievement in differing science and demographical areas. *Journal of Computers in Mathematics and Science Teaching* 18(2): 135–144.

Christmann, E. P., J. L. Badgett, and R. Lucking. 1997. The effectiveness of microcomputer-based computer-assisted instruction on differing subject areas: A statistical deduction. *Journal of Educational Computing Research* 16(3): 281–296.

Christmann, E. P., R. A. Lucking, and J. L. Badgett. 1997. The effectiveness of computer-assisted instruction on the academic achievement of secondary students: A meta-analytic comparison between urban, suburban, and rural educational settings. *Computers in the Schools* 13(3/4): 31–39.

Mauriel, J. 1989. Strategic leadership for schools. San Francisco: Jossey-Bass.

Mayes, R. 1992. The effects of using software tools on mathematics problem solving in secondary school. *School Science and Mathematics* 92(5): 243–248.

Roblyer, M. 1985. Measuring the impact of computers in instruction: A non-technical review of research for educators. Washington, DC: Association for Educational Data Systems.

Watson, P. 1972. Using the computer in education. Englewood, NJ: Educational Technology Publications.

About the Authors

Edwin P. Christmann is a professor, chairman of the secondary education department, and graduate coordinator of the mathematics and science teaching program at Slippery Rock University in Slippery Rock, Pennsylvania.

Roxanne R. Christmann is a certified elementary and special education teacher in Pennsylvania.

John Harrell is a chemistry teacher at North Allegheny High School in Wexford, Pennsylvania.

Adam J. Holy is a graduate student at Slippery Rock University in Slippery Rock, Pennsylvania.

Jill Konton is a biology teacher at North Allegheny High School in Wexford, Pennsylvania.

Jeffrey Lehman is a professor of the Secondary Education/Foundations of Education Department at Slippery Rock University in Slippery Rock, Pennsylvania.

Lyndsay B. Link is a graduate student at Slippery Rock University and teaches chemistry at Butler High School in Butler, Pennsylvania.

Robert A. Lucking is a professor and chairman of the curriculum and instruction department at Old Dominion University in Norfolk, Virginia.

Justin Sickles is a certified chemistry teacher in Pennsylvania.

Mervyn J. Wighting is an assistant professor of education at Regent University in Virginia Beach, Virginia.

Index

Index

Index

X

Y

Z